Reporting Humanitarian Disasters in a Social Media Age

AF705272

From the tsunami to Hurricane Sandy, the Nepal earthquake to Syrian refugees—defining images and accounts of humanitarian crises are now often created not by journalists but by ordinary citizens using Twitter, Facebook, YouTube, Instagram and Snapchat. But how has the use of this content—and the way it is spread by social media—altered the rituals around disaster reporting, the close, if not symbiotic, relationship between journalists and aid agencies, and the kind of crises that are covered? Drawing on more than 100 in-depth interviews with journalists and aid agency press officers, participant observations at the Guardian, BBC and Save the Children UK, as well as the ordinary people who created the words and pictures that framed these disasters, this book reveals how humanitarian disasters are covered in the 21st century—and the potential consequences for those who posted a tweet, a video or photo, without ever realising how far it would go.

Dr Glenda Cooper is a lecturer in journalism at City, University of London. Her research centres on humanitarian disasters, the relationship between aid agencies and journalists and the ethical issues surrounding use of user-generated content. She is the co-editor of *Humanitarianism, Communications and Change* (Peter Lang, 2015), and editor of *The Future of Humanitarian Reporting* (City University, 2014). Before that, she was a Guardian Research Fellow at Nuffield College, Oxford, and a staff journalist at the BBC, *Independent, Daily Mail, Washington Post, Sunday Times* and the *Daily Telegraph*.

Routledge Research in Journalism

News Across Media
Production, Distribution and Cosumption
Edited by Jakob Linaa Jensen, Mette Mortensen, and Jacob Ørmen

Pursuing an Ethic of Empathy in Journalism
Janet D. Blank-Libra

Journalistic Role Performance
Concepts, Contexts, and Methods
Edited by Claudia Mellado, Lea Hellmueller and Wolfgang Donsbach

News of Baltimore
Race, Rage and the City
Edited by Linda Steiner and Silvio Waisbord

The Trump Presidency, Journalism, and Democracy
Edited by Robert E. Gutsche, Jr.

Russia's Liberal Media
Handcuffed but Free
Vera Slavtcheva-Petkova

Critical Perspectives on Journalistic Beliefs and Actions
Global Experiences
Edited by Eric Freedman, Robyn S. Goodman, and Elanie Steyn

Economic News
Informing the Inattentive Audience
Arjen van Dalen, Helle Svensson, Antonis Kalogeropoulos, Erik Albæk and Claes H. de Vreese

Reporting Humanitarian Disasters in a Social Media Age
Glenda Cooper

For a full list of titles in this series, please visit www.routledge.com

Reporting Humanitarian Disasters in a Social Media Age

Glenda Cooper

Routledge
Taylor & Francis Group
NEW YORK AND LONDON

First published 2019
by Routledge
52 Vanderbilt Avenue, New York, NY 10017

and by Routledge
2 Park Square, Milton Park, Abingdon, Oxon OX14 4RN

First issued in paperback 2020

Routledge is an imprint of the Taylor & Francis Group, an informa business

© 2019 Taylor & Francis

The right of Glenda Cooper to be identified as author of this work has been asserted by her in accordance with sections 77 and 78 of the Copyright, Designs and Patents Act 1988.

All rights reserved. No part of this book may be reprinted or reproduced or utilised in any form or by any electronic, mechanical, or other means, now known or hereafter invented, including photocopying and recording, or in any information storage or retrieval system, without permission in writing from the publishers.

Trademark notice: Product or corporate names may be trademarks or registered trademarks, and are used only for identification and explanation without intent to infringe.

Library of Congress Cataloging-in-Publication Data
A catalog record for this book has been requested

ISBN 13: 978-0-367-58419-1 (pbk)
ISBN 13: 978-1-138-48357-6 (hbk)

Typeset in Sabon
by Apex CoVantage, LLC

Contents

Illustrations

Preface and Acknowledgements

Some stories always stay with you as a journalist. In 2004, I was working for the *Sunday Times* in London when the Indian Ocean tsunami happened. As a result of the way I saw how ordinary people's content was used—and the way aid agencies reacted to the disaster—I began to question how journalism was changing. In subsequent years I went to become the Guardian Research Fellow at Nuffield College, Oxford, and then to do a doctorate at City, University of London looking at the use of social media and user-generated content and how this affected the relationship between aid agencies and journalists as a result. But the genesis of this book can be traced back to those first days when I saw picture desks sending their photographers not just to the scene of the disaster, but to airports to pick up video footage shot by returning tourists, changing the visual tropes of how we report such disasters for ever.

Nearly 15 years on, camcorders have been replaced by smartphones; emails replaced by Facebook Live and Instagram stories. Yet while the technology may have changed, the ethical and journalistic questions are still as pertinent as ever.

This research is a snapshot in a rapidly changing media environment; many of these interviews were conducted during 2013–15. But this does not undermine its conclusions. Rather it acts as a significant starting point in order to measure the matters that still are up for debate. How do we treat those caught up in a traumatic situation and who share their content without knowing how far it may travel? How has the collapse of the financial model of journalism impacted the close relationship between journalists and aid agencies? What is the significance of aid agencies increasingly employing former journalists in their press offices? And what problems can the use of bloggers and social media influencers bring to aid agencies?

More than 100 people agreed to be interviewed for this research—journalists, aid agency press officers and creators of user-generated content. This book would not have been possible without the generosity of the time that they gave. The insights, and often frank reflections from journalists on their craft and aid agencies on their work helped to shape it. I was also particularly fortunate enough to secure the co-operation of many UGC creators whose content had been used (sometimes unknowingly until I contacted them); my heartfelt thanks also go to them for their time and honesty. I hope I have repaid their trust by writing up a fair and respectful account.

The journalists and aid agency press officers were from organisations with a base in the UK. The UK has a particularly competitive media and NGO base; and as such can be a useful case study for how disaster reporting is changing. The journalists came from both a print background (quality, mid-market and tabloid), and broadcast organisations (public sector and commercial). The aid agencies that I spoke to were those who had the resources and contacts to engage in regular media work, as members of the Disasters Emergency Committee—the umbrella organisation under which major agencies in the UK join forces when there is a significant acute disaster. The user-generated content creators were selected by looking at liveblogs produced by the *Guardian* and BBC as the Great East Japan quake happened, and identifying content and its creators that was used in those blogs. Further UGC creators I interviewed whose content was not used on those particular liveblogs were selected on the basis of recommendation by journalists whom I interviewed for the microethnographies and in two cases by other UGC creators.

This leads me to further acknowledgments. *The Guardian*, the BBC and Save the Children UK in particular were generous in allowing me to come and sit in their offices and observe what they do, and particular thanks as well to those who facilitated my entry into those places.

I am also very grateful to City, University of London for providing a stimulating academic base to develop this work, and in particular my doctoral supervisor Prof Howard Tumber for his support and guidance. Thanks also to Prof Suzanne Franks, head of department, Dr Mel Bunce and the Humanitarian News Research Network, and Prof Simon Cottle at JOMEC.

Thanks also to those who read this work at different stages and gave excellent advice and to those who have always offered support. In particular, thank you to Dr Judith Townend, Dr Richard Danbury, Dr Kate Wright, Dr Martin Scott, Liz Hunt, Prof Steven Barnett, Prof Karin Wahl-Jorgensen, Dr Josephine Machon and Jessica Cornish.

And finally, as always, so much gratitude to my husband Dominic, who always believed in this book—and to Faith and Alethea who put up with me writing it.

1 Introduction

September 1943. Famine engulfed Calcutta in what Ian Stephens, then editor of the Indian publication *The Statesman*, described as "slow dispirited noiseless apathy" (Stephens, 1966:184). Absorbed in running his newspaper and with limited staff, Stephens was one of those who initially failed to notice what was going on. But when he—and others—did start to comprehend the scale of the unfolding tragedy, he found official denials and cover-ups all around him.

When Stephens broke ranks and started to report on the famine, he was warned by the British government's chief press adviser, B.J. Kirchner, that the newspaper had come "very near the borderline and might get into trouble" (Stephens, 1966:185). Cables leaving Bengal were watched carefully. All "such meaningful words as famine, corpse, starvation were methodically struck out" by India's central government (Stephens, 1966:187) and the Bengal authorities stopped publishing mortality statistics. But, for two months, *The Statesman* waged a campaign to expose the truth about the famine. Even so, estimates are that between seven and ten million people died of starvation, malnutrition and disease during 1943–44, out of a population of 60 million—and for too long, the wider world had simply no idea what was going on because the information was suppressed.

January 2010. An earthquake measuring 7.3 on the Richter scale hits Haiti at 16.53 local time on January 12. At 17.00, a Haitian—Fredo Dupoux—tweeted: "*Oh shiet [sic] heavy earthquake right now! In Haiti*"; another tweeter @FutureHaiti said "Earthquake 7 Richter scale just happening #Haiti" (Macleod, 2010). Figures vary on how many people died or were displaced,[1] but the eyes of the world were on the disaster within minutes. As the quake and its aftermath unfolded, according to the Twitter-tracking service Sysomos, 2.3 million tweets included the words 'Haiti' or 'Red Cross' between 12 and 14 January (Evans, 2010).[2]

November 2017. The eruption of Mt Agung in Bali leads to more than 100,000 Balinese being evacuated from their homes. The international airport is closed. *Time* and the *Daily Telegraph* report that tourists are flocking to Instagram to take beautifully constructed selfies of themselves in front of the ash cloud (Abrams, 2017; Smith, 2017), the latest manifestation of the 'disaster selfie' previously seen at the scenes of the 2015 Nepal quake and the Grenfell Tower disaster (Ibrahim, 2015; Meade, 2017).

On the surface, the contrast between the cover-up of the Bengal famine in 1943 and the extensive coverage of the Haiti earthquake nearly 70 years later or the Bali eruption could not appear more stark: one characterised by no information and disinformation; the other by 'ordinary' people describing the disaster to anyone who cared to listen around the world as it happened. In Haiti, even aid workers and journalists initially played catch-up with the public on the ground, who were alerting them to where the problems were and where aid was needed.

It is no wonder that the journalist Anand Giridharadas wrote of the experience:

> The old paradigm was one-to-many: foreign journalists and aid workers jet in, report on a calamity and dispense aid with whatever data they have. The new paradigm is many-to-many-to-many: victims supply on-the-ground data; a self-organizing mob of global volunteers translates text messages and helps to orchestrate relief; journalists and aid workers use the data to target the response.
>
> (Giridharadas, 2010:3)

The Haitian earthquake came six years after the 2004 Indian Ocean tsunami, seen as a turning point (Gillmor, 2005)—where images and accounts taken by ordinary people superseded those taken by journalists. In the tsunami's aftermath, the claims made for user-generated content (UGC) were extensive: that it would transform reporting, making a more diverse range of stories and voices heard; that creators of content could be active shapers rather than passive bystanders in their own stories (Gillmor, 2005; Glocer, 2006; Jenkins, 2006; Deuze, 2007). And no longer would the world be able to turn its back on those suffering in faraway places: this could be a revolution in reporting. Or so UGC's early supporters thought.

The Aim of This Book

As a journalist working at the UK's *Sunday Times* at the time of the 2004 tsunami[3] it soon became clear that this disaster was being covered in a very different way. While we were, of course, trying to get our own journalists out there as soon as possible, I found that the use of camcorder footage from tourists on the beaches, as well as blogs such as waveofdestruction.org sharing tsunami photos, seemed to be ushering in a new era of what some scholars called convergent and collaborative journalism. A paradigm shift appeared to be taking place "in which once the media was the centre of the universe and now the user is the centre of the universe" (Robinson and De Shano, 2011:977).

The impact it had on aid and aid workers was equally groundbreaking with the coverage affecting the scale of the donations. In the newsroom, looking for new angles on the tsunami story, on 4 January 2005, I came across a Press Association story, saying the charity Medecins Sans Frontieres issued an unprecedented statement asking that no more money be given to its fund earmarked for the tsunami. The story spread across the world (Bennhold, 2005). Aid agencies asking *not* to be given money? It seemed incredible—and indeed it was a statement that was attacked by other aid agencies at the time but which proved to be judicious (Barbagallo, 2005).

A second development was the 'DIY aid squad' (Hodson, 2005) which focused on so-called 'guerrilla aid workers' who had watched the coverage on TV, and were determined to go and help themselves. This of course again was not new in itself—there had been previous examples during 1990s crises in particular the Balkans war and the collapse of Romania, but these aid workers in the 2000s were able to use blogs and film to publish what they were doing (Campbell, 2005; Hartley-Parkinson, 2005; Thompson, 2007; Ramos and Piper, 2005).

Six months later, journalists were confronted by blogs and images, which had been created by the survivors and witnesses to the 7/7 terrorist attacks when journalists themselves could not get near the sites.[4] It became clear that the tsunami was only the beginning: ordinary people began to share their images and texts of elections, conflicts, disasters and riots.[5] By a decade later such was the ubiquity of social media as a communications tool that there would be a US president who chose to communicate messages to his electorate and fellow world leaders not via official bulletins or interviews with the press but over Twitter, whether announcing airstrikes or conducting diplomacy with the North Koreans.[6]

Yet 15 years on from the 2004 tsunami, those camcorder footage and blogs now appear to be ancient history. The availability of the smartphone to capture stills and video, coupled with the growth of social networking sites such as Facebook, Twitter, Instagram and Snapchat, as well as the ability to livestream via platforms such Facebook Live or Periscope, have made publishing and sharing content far easier for the public.

However, the early optimism of a new era of collaboration and equalisation, in which different stories would be told and different voices heard has not necessarily been borne out, as the Bali volcano showed.

This book examines the effect the increasing availability of words and images produced by non-journalists ('user-generated content') and facilitated by social media platforms has had on the reporting of humanitarian disasters and traditional source–media relationships between journalists and aid agencies. But rather than focusing on textual analysis as many researchers have done (Naab and Sehl, 2017), this book takes an empirical approach and speaks to journalists, aid agencies and creators of UGC.

As a result, I interviewed more than a hundred people, mainly face to face, although for those who whom this was impossible sometimes via Skype or email. The idea was to look at the informal reality, which can only be perceived from the inside—whether that is from inside media organisations or within the NGOs that are their sources. It allowed the processes that lead to the coverage we the audience receive, and on occasion create, to be examined: how and why UGC is used, what relationship creators of UGC have with the media, and what procedures NGOs use when dealing with the media.

The subjects of the interviews were, in the main, 'elite' (Gillham, 2000)—journalists and NGOs who have particular experience in this field. However, there were also a considerable number of 'accidental journalists'—members of the public who found themselves caught up in humanitarian crises (Allan, 2013).

Humanitarian Disasters—Definitions

The definition of 'humanitarian disaster' has traditionally proved contentious, whether it is 'natural' or 'man-made', and what scale of destruction qualifies it as a disaster.

This book looks mainly at crises and disasters that fall under the contentious heading of 'natural', but will use the term 'humanitarian' because of the problems with the terminology 'natural' and to emphasise that the disasters I will be looking at have had an impact on human societies.

In 2017, 318 'natural' disasters occurred across the world, killing 9,503 people and affecting more than 96 million others; according to the Centre for Research of the Epidemiology of Disasters (CRED), a World Health Organization collaborating centre.[7] While the mortality figures were considerably lower than the average for 2007–16 of 68,302 deaths yearly, the 2017 disasters caused significantly more economic damage worth US$314 billion, reflected in the impact of three hurricanes—Harvey (US$95 billion), Irma (US$66 billion) and Maria (US$69 billion), affecting the United States and the Caribbean (Wallemacq, 2018). The low mortality rate for 2017 is explained by the fact that there was no single major event responsible for increased mortality. This is unlike more recent years where the earthquake in Nepal (2015) killed 8,831 people and Typhoon Haiyan in the Philippines (2013) killed 7,354.

While hurricanes, floods and earthquakes are the result of what we might see as weather events or geological faults, Cottle points out that they are "increasingly known as '*un*natural phenomenon'" (2009:43), while Beck calls them the products of the "manufactured uncertainties" of the global risk society (1999:19).

For the scale of a disaster—or even the fact a 'natural' event becomes a disaster all—is often the result of the existing living conditions of the people caught up in it. For example, the Guatemalan earthquake of 1976 affected the slum dwellers of Guatemala City and the Mayan Indians living in impoverished towns so much more than their more affluent neighbours that it was eventually dubbed a "classquake" (Wisner et al, 2006). This explains, too, why Haiti's 2010 earthquake (measuring 7.0 on the Richter scale) was much more devastating than the 2010 and 2011 Canterbury and Christchurch earthquakes in New Zealand (7.1 and 6.3 on the Richter scale).[8] Haiti's frail infrastructure and food insecurity meant the quake effectively destroyed a country already on the brink of collapse (Kolbe et al, 2010).

Food crises are also far from 'natural' phenomena. While drought affects crop growth, famines occur because of political situations (see Sen, 1999). The rise in 11 million people facing food insecurity and requiring urgent humanitarian action between 2016 and 2017 was because of new or intensified conflict in Myanmar, north-east Nigeria, Democratic Republic of Congo, South Sudan and Yemen (Food Security Information Network, 2018). This can lead to incongruous situations: in the food crisis occurring in Niger in 2005, there were markets full of food 2km away from the place where the BBC were filming stories that would lead the Ten O'Clock News (Vasagar, 2005).

Both the International Federation of the Red Cross and Red Crescent Societies (IFRC) and CRED define 'natural disasters' as ones where there is disruption of a community or society's functioning, and losses which exceed that community or society's ability to cope using only its own resources and end up by needing assistance from a national or international level. For a disaster to be entered into the CRED database, at least one of the following fairly modest criteria of scale must be fulfilled:

- ten or more people reported killed
- 100 or more people reported affected
- declaration of a state of emergency
- call for international assistance.[9]

What these definitions have in common is that they describe events that affect *human* society, rather than that happen in deserted areas, i.e. "extreme natural events are not disasters until a vulnerable group of people is exposed" (Wisner et al, 2006:i).

The link between 'natural' events and their consequences for human beings can be expressed in another way—by looking at how such events are anticipated, prepared for and responded to. So-called 'disasters' tend to occur when they are not prepared for properly, or there are not sufficient social structures to cope with them when they occur; such 'natural' disasters are in fact symptoms of incompetency or irrationality in managing resources (Cottle, 2009; see also Benthall, 1993). Rogers (1978) points out one further important factor in defining disasters—that they are events in which normal communication channels are disrupted, just when information flows are needed most.[10]

For disasters can be seen as socially produced phenomena (Cottle, 2009), not just because of the lack of infrastructure or strong government that means they affect the poor so much more than the rich, but also because of the way they are covered (or not) by the news media. As Cohen observes:

> Coverage is so selective that the media in effect create a disaster when they recognize it: they give institutional endorsement or attestation to bad events which otherwise have a reality restricted to a local circle of victims.
>
> (Cohen, 2001:169)

Finally, a word about the role that conflict plays in disasters and the history of how conflict has been reported. As mentioned earlier, one of the issues that can exacerbate a disaster is clearly conflict (Benthall, 1993:12)—for example, the 2011 food crisis in the Horn of Africa was undoubtedly made worse in Somalia because of the ongoing conflict there (Hillier and Dempsey, 2012). Yet unlike disaster reporting, war reporting has been treated by the mainstream media as a strand of reporting in itself. In the past (at least), there tended to be reporters who specifically identified themselves as war reporters (see Bell, 1995; Marinovich and Silva, 2000; Bowen, 2006; even, reluctantly and humorously, Ayres, 2005). However, apart from very few examples, such as David Loyn and, before him, George Alagiah (both former BBC developing world correspondents), reporters do not tend to specialise in disaster reporting.

So the journalists assigned to disasters are either those in local bureaux or the so-called 'fireman', 'parachute' or 'special' correspondents—senior heavyweight reporters who are flown in to cover 'big' stories, rather than covering such issues full-time. While they are very experienced, they are not dedicated to this type of journalism. And so, compared with conflict, there is less of an established academic pathway for looking at humanitarian disasters.

Humanitarian Crises, UGC and Social Media

UGC has transformed all kinds of reporting, but it has proved particularly popular in reporting humanitarian crises because of the fact that the mainstream media has faced declining revenues, culminating in budget cuts that saw traditional foreign bureaux closed or their funding slashed (Harding, 2009; Brook, 2008, 2009; Sambrook, 2010). Foreign reporting is expensive (Carroll, 2007). Covering disasters in places where infrastructure has broken down is *very* expensive—and can often take considerable time to get to inaccessible places. How seductive is it, therefore, when a news editor still trying to get a correspondent on a plane suddenly sees graphic accounts and visuals of a landslide, an earthquake a fire or a flood on her Twitter feed, Facebook page or via Instagram stories?

But there are well-documented problems associated with using UGC. How accurate are the reports and how can they be verified? (Hermida, 2012; Hayward, 2013; Newman, 2009; Schifferes et al, 2013). Who are the authors of this content and what preconceptions and misconceptions do they have (Paulussen and Ugille, 2008; Bruns, 2009:129)?[11] For being part of the action can also limit the story told; UGC can accurately represent the situation an individual finds themselves in, but can also show only a partial version of events (Seaton, 2005).

And UGC may not even promote alternative voices. An analysis of the BBC's use of user-generated content in 2008 found a typical contributor was a white male aged 55–59, employed full-time as a non-manual worker (Wardle and Williams, 2008). Certainly, the Instagram posts from the Bali volcano in 2017 shared to a wider audience via the developed world media were not from native Balinese who were having to flee their homes. Instead they were of tourists who showed off their best yoga poses and bikini shots in front of the ash cloud (Smith, 2017). Far from elucidating and authenticating this crisis, the purpose of these 'disaster selfies' and others taken in the aftermath of catastrophic events was not to "extend the project of empathy or solidarity with the fallen or the incapacitated but. . . [are] proclaiming a statement of survival against a backdrop of disaster or carnage" (Ibrahim, 2015:217).

UGC may also skew the definition of what is important towards the unexpected or the spectacular. It often lends itself to the dramatic over the chronic, the earthquake over the long-term famine. It is not an accident that the disasters we have primarily seen framed through the lens of UGC are ones like Hurricane Irma in 2017 rather than the food crisis in the Sahel. As one journalist put it: "The problem with citizen journalists—just like all us citizens—is that they're incorrigible sensationalists" (Sutcliffe, 2007:28).

If there is verifiable UGC available, it is also important to consider how the mainstream media treats the creators of such content. Is permission sought from, and the privacy respected of, people who may have been caught up in the most traumatic of circumstances (Joye, 2013; Duncan and Newton, 2014; Townend, 2012)? Do journalists, who have had to rethink the way they approach witnesses of terrible events when they encounter them face to face or over the phone, use the same sensitivity when approaching creators of content via email or social media (Newton and Duncan, 2012)?

Meanwhile, the relationship between journalist and UGC creator in humanitarian disaster reporting has not been the only one to change. The use of such content has also forced a re-examination of the symbiotic relationship between non-governmental organisations (NGOs) and the mainstream media.

Carlson (2009) builds on Gans (1980) to extend the metaphor of the source–journalist relationship as a duel, dance or fight for domination. Certainly, aid agencies—as traditionally one of the most important sources in humanitarian reporting—have potentially had to rethink how they approach this dance to maintain their influence over the way disasters are depicted. They have responded to the challenge of user-generated content by professionalising and taking an increasingly active role in trying to frame the media agenda around disasters (Powers, 2015a; Abbott, 2015; Wright, 2015, 2018; Waisbord, 2011; Beckett, 2009). They have done this by transforming their press offices into pseudo-news operations, and by using social media to run online campaigns via Twitter and Facebook.

Some of the biggest aid agencies in the field have shifted towards pro-active advocacy as well as aid delivery. They are now more than simple providers—they are working as political pressure groups or as policy institutes (Powers, 2015a, 2015b). This changes their message and their relationship with the media, with advocacy now a key part of their work.

Yet, aid agencies are also potentially becoming embroiled in Altheide and Snow's (1979) idea of "media logic", which Cottle argues is far removed from the ideals and aims of humanitarianism (Cottle and Nolan, 2009). And, while aid agencies may be perceived as closer to Anderson's (1993, 1997) 'influential outsiders' than Hall et al's (1978) 'primary definers', it would be wrong to see them as outside the corridors of power. For example, the 'insider–outsider' approach adopted during Make Poverty History in 2005 led one critic to accuse NGOs of handing the G8 the "wristband of legitimacy" (Burgis, 2005); one of Tony Blair and Gordon Brown's closest advisors, Justin Forsyth, went on to become chief executive of Save the Children UK (Cartmell, 2010).[12]

The Structure of the Book

To analyse these points, the book is structured as follows. After this introduction, the next chapter sets up the theoretical background for the book, examining recent literature around coverage of humanitarian disasters, in particular the ritualisation of humanitarian coverage, discussion of 'distant suffering', the CNN effect and the use of field theory in a UGC context.

Digital technologies are altering the way we can get access to news and create different dynamics in the way news is produced and published (Tumber, 2014). Methods and production of journalism are changing in a world of new media technologies. So by using Bourdieu's field theory I explain and understand media concentration, personalisation of the media and the changes wrought by the fact that today anyone can self-broadcast/publish and the effect this has on journalism.

This builds on previous sociological work (e.g. Tuchman, 1972; Tunstall, 1974), which looked at the importance of journalists' routines, practices and power relationships in the past. In trying to understand the upheaval that journalism is going through and the new structures and practices used for producing news, Habermas's public sphere and Bourdieu's field theory prove a useful theoretical framework.

The chapters that follow look at the role of journalists, aid workers and UGC creators, whose content was used knowingly or unknowingly by the mainstream media. It also includes microethnographies at two media organisations and an aid agency, as well as analysis of mainstream media liveblogs on the Great East Japan quake of 2011.

The chapters examine how journalists and aid agencies understand their relationship and control of the story. The involvement of UGC creators and their content in humanitarian reporting has, it is claimed, led to a potential disruption in the close and symbiotic relationship between journalists and aid agencies (Cottle and Nolan, 2009; Cooper, 2007, 2011a; Wright, 2015). It therefore assesses the state of play in this association—described memorably by one of the interviewees as "a mucky weekend not a one-night stand". It examines boundary (re)negotiations, in particular how both groups describe working together in the field, and the use of aid agency content and compares and contrasts their answers.

It then analyses how journalists and aid agencies see the use of UGC. Speaking to some of the leading journalists in the UK field, it asks a series of questions about how UGC has changed their working conditions and how they make use of it; the major aid agencies in the UK are asked similar questions.

It lays out the importance that journalists and NGOs attach to user-generated content, explaining how both use it and how they conceptualise the creators in a way that justifies their own processes. In particular it looks at how these two groups articulate the issues of verification, privacy, consent, copyright and graphic images involved in the use of UGC in the newsroom and press office.

It goes on to detail the way aid agencies have grappled with social media and UGC and found it difficult to integrate successfully. They have responded to the challenge of user-generated content by professionalising and taking an increasingly active role in trying to frame the media agenda around disasters (Powers, 2015a; Abbott, 2015; Wright, 2015; Waisbord, 2011; Beckett, 2009). They have done this by transforming their press offices into pseudo-news operations, and by using social media to run online campaigns via Twitter and Facebook. It looks in detail at the use of bloggers and Instagram influencers as a halfway house by aid agencies such as Plan UK, World Vision UK and Save the Children, and also goes behind the scenes at the ambitious Save the Children Global Media Unit which was set up to co-ordinate and control the international agency's media messaging on a 24-hour basis around the world.

In comparison, the way journalists have tried to take control of UGC is examined through two case studies—that of GuardianWitness and the BBC's UGC Hub to assess how mainstream media organisations have sought to control the threat posed by UGC by adopting its tropes, and assimilating the content itself.

This chapter discusses how the mainstream media has sought to capture and use user-generated content, taking ownership of it in a more organised way than simply mimicking typical UGC tropes. Just as the media sought to 'normalise' blogs (Singer, 2005), now media organisations have turned to popular forms of UGC, mainly pictures. But to maintain their authoritative position, they have used strategies to make sure such content is segregated or co-opted (Wahl-Jorgensen, 2015).

Thus, the BBC UGC Hub successfully promoted the idea of user-generated content as another source for the corporation's journalists, rather than a different (and potentially challenging) form of journalism altogether. The Hub's journalists had evolved to act as effective 'gatewatchers' (Bruns, 2008) and gatekeepers, both requesting content, via the 'Have Your Say' form, and choosing it, when there was breaking news. By setting up the Hub, the BBC was taking ownership of user-generated content, and the corporation's journalists mostly experienced UGC through the medium of the Hub. With the GuardianWitness app, journalists spoke of their dedication to giving opportunities to different voices and to telling different stories.

Yet, the overall control of the stories and the pictures was something the staff found difficult to justify. There were also problems in reaching out to those at the centre of these stories. As a result, the content was more likely to be dominated by NGOs with their own particular agendas to explain than those who had experienced a crisis themselves; this was something that staff at GuardianWitness co-operated with and accepted in order to get the content they required.

A key aspect that is often overlooked in UGC studies is understanding of the creators themselves. The book then goes on to use the case study of the 2011 Japanese quake to look at how user-generated content was used by two mainstream media organisations—and how those whose content was used felt about it. It discusses whether permission was sought for the use of the content and whether consent could be meaningfully given, given the traumatic situation those creating content were caught up in. It looks specifically at the *Guardian* and BBC liveblogs of the 2011 Great East Japan quake to analyse how much of this UGC comes from the 'diverse voices' that early researchers envisaged, and how much content via social media comes from opinion formers, authority figures and other media.

In particular I introduce the concept of the '*virtual doorstep*'—how creators of UGC have found themselves hounded online by journalists and the consequences it had for them. Through 30 interviews with UGC creators, I analyse how the mainstream media treated them and how they dealt with this, with lessons for the future outside the boundaries of humanitarian disasters such as the 2017 Manchester attack. In the main, the creators did not know their content was going to be used and while they were often flattered and pleased to see their tweets and images reach a wider audience, they also faced problems such as trolling, their personal and professional identities colliding as content was spread far further than they ever guessed or intended. For some, the only response was to erase their social media identities and/or content either temporarily or permanently.

Given the unresolved issues that many of the UGC creators articulated, the book then examines the ethical questions in a UGC world. Coverage of humanitarian disasters has always had complex ethical implications—acting as a moral "bellwether" (Moeller, 1999) signalling normative responses on behalf of the audience. But in a media environment, which is challenging our understanding of who can be a journalist, and how content can be easily shared and appropriated in ways far beyond what was originally intended, the ethics of reporting humanitarian disasters has to be addressed. Ward (2005, 2008) argues that the globalisation of news media requires a rethinking of journalism ethics to a more 'humanistic' approach, a point taken up by Wahl-Jorgensen and Pantti who talk of the need of the journalist to be "actively involved in shaping and constructing the meaning of particular events" (2013:193) in order that ethical responsibilities are formed both at global, national and individual levels.

So the next chapter focuses particularly on whether those who create content are treated respectfully in terms of copyright, credit and having their privacy respected. The mainstream media has often taken a smash-and-grab approach to UGC—trawling Facebook, Twitter and YouTube for content, often without thought for the consequences. But this raises important issues around the rights and responsibilities for the people who create it and the media outlets that distribute more widely—often more widely than the creator ever thought possible. Most previous literature relates to face-to-face interviews, but in this chapter I would argue it is equally applicable to online encounters as well—something as yet not addressed adequately by journalistic codes.

Finally, the conclusion sums up the empirical and theoretical discussions in the book in particular the blurred boundaries and (re)negotiations that are taking place in the world of humanitarian reporting. It details how the source–media relationship has changed and the growth of the 'virtual doorstep'. It looks at how some creators have tried to reframe the narrative and (to coin a phrase) take back control. It discusses the ramifications of these new approaches to humanitarian reporting and whether they have made a significant difference.

It looks again at the concept of distant suffering and whether the "anaesthetizing effects" of traditional disaster reporting (Chouliaraki, 2006:25) have been overcome by these new forms. It considers whether UGC creators have disrupted the journalistic field and, if so, how journalists and aid agencies have reacted to contain and control this disruption. It ends by looking at how the knowledge gained from examining humanitarian stories can be widened out for crisis news stories more widely.

The explosion in user-generated content in recent years made this book a potentially massive undertaking. To keep some kind of focus, therefore, I limited myself to British newspapers and broadcast channels with a substantial presence in the UK (the BBC, ITV, Sky News, Channel 4 News and Al Jazeera English). I also limited myself to UK-based aid agencies, principally those who are part of the umbrella organisation, the Disasters Emergency Committee (DEC). The UK has a particularly competitive media market—and was the birthplace of many of today's most influential INGOs but the conclusions that I draw could be pertinent to many other developed world countries.

As the CRED statistics show, humanitarian disasters remain a constant in our world. To understand how we make sense of them as victims, journalists, NGOs and audiences we have to understand how and why they have been reported as they have been in the past.

For example, the problems around using Instagram influencers were not dissimilar from those aid agencies who had earlier used blogs to try to publicise stories; the problems that aid agencies wrestled with over livestreaming platforms such as Facebook Live echoed earlier conversations about the earlier relevance of using Twitter.

But the need for access to such disaster zones by journalists, the need for publicity for aid agencies and the determination of both to keep control of the story in the face of a disrupted field thanks to the emergence of widespread UGC from ordinary people and social media platforms remained the same whether it was the 2004 tsunami they were talking about, the 2014 typhoon in the Philippines or the 2018 Guatemalan volcano eruption. And getting it right is vital for all three groups. As aid agencies know, communication is a form of aid (Ghannam, 2016). As journalists realise despite the concerns over compassion fatigue (Moeller, 1999), people will still react to these stories. And for those caught up in a crisis beyond their ability to cope, the growth of UGC means that they potentially have a voice that they can use which challenges the symbiotic relationship that previously characterised such narratives.

Notes

1. The official death toll according to the Haitian government was 316,000, but this is widely disputed. Other suggestions ranged from 158,000 to as low as 46,000 (O'Connor, 2012; Kolbe et al, 2010; BBC, 2011).
2. There were also 189,024 tweets mentioning '90999'—this was a number that could be used to text message a donation to the Red Cross via mobile phone.
3. I was working as deputy editor at the *Sunday Times News Review* from November 2004 to May 2005, and commissioned and edited pieces in the aftermath of the tsunami.
4. In particular, journalists drew on the urban 75 blogs (www.urban75.org/london/bombs.html) and the anonymous blogger Rachel North London (http://rachelnorthlondon.blogspot.com/2005/07/kings-cross-bomb-my-eyewitness-account.html) who helped set up the survivors' group Kings Cross United (Wall and El Zahed, 2015).
5. These include the 7/7 bombings (Allan, 2006; Douglas, 2006; Sambrook, 2005); Hurricane Katrina (Robinson, 2009); the Sichuan earthquake (Wang, 2010; Nip, 2009); the Mumbai terror attacks (Dolnick, 2008, Belair-Gagnon, 2012); the 2009 Iran elections, in particular the death of Neda Agha-Soltan (Mortensen, 2011; Burns and Eltham, 2009; Fathi, 2009); the 2010 Haiti earthquake (Pantti, Wahl-Jorgensen and Cottle, 2012; Chouliaraki, 2013); the Arab Spring revolts and Syrian conflicts (Allan, 2013; Wardle, Dubberley and Brown, 2014; Wall and El Zahed, 2015; Johnston, 2015); the 2011 Great East Japan quake and

tsunami (Acar and Muraki, 2011; Hjorth and Kim, 2011; Utz, Schultz and Glocka, 2013); the 2011 London riots (Baker, 2011, 2012; Lewis et al, 2011); the Boston Marathon bombings (Manjoo, 2013; Suran and Kilgo, 2015); and the Paris attacks (Cassidy, 2015; Frankel, 2015).

6. See https://twitter.com/realDonaldTrump/status/984022625440747520 and https://twitter.com/realdonaldtrump/status/948355557022420992?lang=en for example.
7. CRED (www.cred.be) was founded in 1973 by Prof Michel F. Lechat, an epidemiologist at the Université Catholique de Louvain. The centre was built on a previous research programme he had set up at the university to study health issues in disaster situations. It became a WHO Collaborating Centre in 1980.
8. Another earthquake in Christchurch took place on 14 February 2016. This was originally reported as 5.9 on the Richter scale but was later amended to 5.7.
9. See www.emdat.be/explanatory-notes.
10. See Morrow et al (2011) and Wall and Chery (2011) for how innovative use of SMS, Facebook, crisis mapping began to alter this in the Haitian earthquake of 2010.
11. Bruns quotes Rusty Foster of the collaborative technology website Kuro5hin as speaking on behalf of a large community of citizen journalists when he notes that "we may be biased, but at least we're obviously biased. And K5 has so many different points of view, that a fairer process can emerge from a balance of biases" (Foster, 2001).
12. Forsyth then went on to Unicef before resigning his post as deputy executive director in 2018 after claims of inappropriate behaviour during his time at Save the Children UK emerged (Greenfield, 2018).

References

Abbott, K. 2015, "How NGOs became the new(s) reporters", in *Humanitarianism, communications and change*, eds. S. Cottle & G. Cooper, Peter Lang, New York, pp. 183–197.

Abrams, A. 2017, 1 December-last update, *Bali tourists flood Instagram with Mount Agung volcano photos as thousands evacuate* [Homepage of Time Magazine], [Online]. Available: http://time.com/5044257/bali-volcano-mount-agung-instagram-posts/ [2018, 8 June].

Acar, A. & Muraki, Y. 2011, "Twitter for crisis communication: Lessons learned from Japan's tsunami disaster", *International Journal of Web Based Communities*, vol. 7, no. 3, pp. 392–402.

Allan, S. 2006, *Online news: Journalism and the internet*, McGraw-Hill Education, Maidenhead.

Allan, S. 2013, *Citizen witnessing: Revisioning journalism in times of crisis*, Polity Press, Cambridge, UK.

Altheide, D.L. & Snow, R.P. 1979, *Media logic*, Sage Publications, Beverly Hills, CA, London.

Anderson, A. 1993, "Source-media relations: The production of the environmental agenda", in *The mass media and environmental issues*, ed. A. Hansen, Leicester University Press, Leicester, pp 51–68.

Anderson, A. 1997, *Media, culture and the environment*, UCL Press, London.

Ayres, C. 2005, *War reporting for cowards*, John Murray, London.

Baker, S.A. 2011, "The mediated crowd: New social media and new forms of rioting", *Sociological Research Online*, vol. 16, no. 4, p. 21.

Baker, S.A. 2012, "From the criminal crowd to the 'mediated crowd': The impact of social media on the 2011 English riots", *Safer Communities*, vol. 11, no. 1, pp. 40–49.
Barbagallo, P. 2005, 1 March-last update, *Healing ways* [Homepage of NonProfit Pro], [Online]. Available: www.nonprofitpro.com/article/though-misunderstood-first-doctors-without-borders-bold-decision-stop-accepting-tsunami-relief-donation-could-help-revive-public-trust-nonprofits-32313/all/ [2018, 8 June].
BBC. 2011, 1 June-last update, *Report challenges Haiti earthquake death toll* [Homepage of BBC], [Online]. Available: www.bbc.co.uk/news/world-us-canada-13606720 [2016, 25 March].
Beck, U. 1999, *World risk society*, Polity Press, Cambridge.
Beckett, C. 2009, *NGOs as gatekeepers to 'Local Media': Networked news for developing countries*, LSE, London.
Belair-Gagnon, V. 2012, "'Getting it Right!': How did social media transform BBC News journalism?" *Communiquer dans un monde de normes. L'information et la communication dans les enjeux contemporains de la 'mondialisation*, p. 235.
Bell, M. 1995, *In harm's way: Reflections of a war zone thug*, 1st edn, Hamilton, London.
Bennhold, K. 2005, 6 January-last update, *Doctors' group refuses more donations: Charity sets off storm with tsunami aid halt* [Homepage of New York Times], [Online]. Available: www.nytimes.com/2005/01/06/news/doctors-group-refuses-more-donations-charity-sets-off-storm-with.html [2018, 8 June].
Benthall, J. 1993, *Disasters, relief and the media*, I B Tauris, London.
Bowen, J. 2006, *War stories*, Simon & Schuster, London.
Brook, S. 2008, 3 September-last update, *Jobs to go in daily mirror cutbacks'* [Homepage of The Guardian], [Online]. Available: www.guardian.co.uk/media/2008/sep/03/mirror.pressandpublishing [2015, 15 August].
Brook, S. 2009a, 5 January-last update, *Telegraph slashes stringer fee to £60* [Homepage of The Guardian], [Online]. Available: www.theguardian.com/media/2009/jan/05/daily-telegraph-media-group-newspapers [2015, 17 August].
Bruns, A. 2008, *3.1. The active audience: Transforming journalism from gatekeeping to gatewatching*. Available: http://snurb.info/files/The%20Active%20Audience.pdf [2018, 19 August]
Bruns, A. 2009, "From reader to writer: citizen journalism as news produsage", in *International handbook of internet research*, eds. J. Hunsinger, L. Klastrup, & M. Allen, Springer, Dordrecht.
Burgis, T. 2005, 3 July-last update, *Insider-outsider: The NGO fracture zone* [Homepage of Open Democracy], [Online]. Available: www.opendemocracy.net/globalization-G8/NGO_power_2643.jsp [2012, 19 May].
Burns, A. & Eltham, B. 2009, "Twitter free Iran: An evaluation of Twitter's role in public diplomacy and information operations in Iran's 2009 election crisis".
Campbell, D. 2005, 14 February-last update, *Tsunami help* [Homepage of Tsunami Help], [Online]. Available: http://dncthailand2005.blogspot.com/ [2018, 8 June].
Carlson, M. 2009, "Dueling, dancing, or dominating? Journalists and their sources", *Sociology Compass*, vol. 3, no. 4, pp. 526–542.
Carroll, J. 2007, March-last update, *Foreign news coverage: The U.S. media's undervalued asset* [Homepage of Joan Shorenstein Center on the Press, Politics and Public Policy], [Online]. Available: https://shorensteincenter.org/wp-content/uploads/2012/03/2007_01_carroll.pdf [2018, 8 June].

Cartmell, M. 2010, 9 July-last update, *Downing street's Justin Forsyth takes chief executive role at save the children* [Homepage of PR Week], [Online]. Available: www.prweek.com/uk/news/1015331/Downing-Streets-Justin-Forsyth-takes-chief-executive-role-Save-Children/?DCMP=ILC-SEARCH [2012, 19 May].

Cassidy, J. 2015, 23 November-last update, *Terrorism in the age of Twitter* [Homepage of New Yorker], [Online]. Available: www.newyorker.com/news/john-cassidy/terrorism-in-the-age-of-twitter [2016, 29 March 2016].

Chouliaraki, L. 2006, *The spectatorship of suffering*, Sage Publications, London.

Chouliaraki, L. 2013, *The ironic spectator: Solidarity in the age of post-humanitarianism*, Polity Press, Cambridge.

Cohen, S. 2001, *States of denial: Knowing about atrocities and suffering*, Polity Press, Cambridge.

Cooper, G. 2007, *Anyone here survived a wave, speak English and got a mobile? Aid agencies, the media and reporting disasters since the tsunami*, Nuffield College, Oxford.

Cooper, G. 2011a, *From their own correspondent? New media and the changes in disaster coverage: Lessons to be learned*, Reuters Institute for the Study of Journalism, Oxford.

Cottle, S. 2009, *Global crisis reporting: Journalism in the global age*, McGraw-Hill Education, Maidenhead.

Cottle, S. & Nolan, D. 2009, 16 November-last update, *How the media's codes and rules influence the way NGOs work* [Homepage of Nieman Journalism Lab], [Online]. Available: www.niemanlab.org/2009/11/simon-cottle-and-david-nolan-how-the-medias-codes-and-rules-influence-the-ways-ngos-work/ [2015, 18 August].

Deuze, M. 2007, "Convergence culture in the creative industries", *International Journal of Cultural Studies*, vol. 10, no. 2, pp. 243–263.

Dolnick, S. 2008, 30 November-last update, *Bloggers provide raw view of the Mumbai attacks* [Homepage of NBC News], [Online]. Available: www.nbcnews.com/id/27984057/ns/technology_and_science-tech_and_gadgets/ [2015, 4 August].

Douglas, T. 2006, 4 July-last update, *How 7/7 'Democratised' the Media'* [Homepage of BBC News], [Online]. Available: http://news.bbc.co.uk/1/hi/uk/5142702.stm [2011, 7 November].

Duncan, S. & Newton, J. 2014, "Emotion and trauma in reporting disaster and tragedy", in *The future of humanitarian reporting*, ed. G. Cooper, City University, London, pp. 69–74.

Evans, M. 2010, 15 January-last update, *Haitian earthquake dominates Twitter* [Homepage of Sysomos], [Online]. Available: http://blog.sysomos.com/2010/01/15/haitian-earthquake-dominates-twitter/ [2012, 17 May].

Fathi, N. 2009, "In a death seen around the world, a symbol of Iranian protests", *New York Times*, vol. 22, no. 6.

Food Security Information Network. 2018, *Global report on food crises 2018*, World Food Programme, Rome, Italy.

Foster, R. 2001, 12 October-last update, *The utter failure of weblogs as journalism* [Homepage of Kuro5hin], [Online]. Available: www.kuro5hin.org/comments/2001/10/11/232538/32?pid=4#5 [2016, 25 March].

Frankel, M. 2015, 20 November-last update, *#Paris: UGC expertise can no longer be a niche newsroom skill* [Homepage of BBC], [Online]. Available: www.bbc.co.uk/

blogs/collegeofjournalism/entries/fbb87059-ab13-4b79-b5ad-5a47a417fb8a [2016, 29 March].
Gans, H.J. 1980, *Deciding what's news: A study of CBS evening news, NBC nightly news, 'Newsweek' and 'Time'*, Constable, London.
Ghannam, J. 2016, *Media as a form of aid in humanitarian crises*, Center for International Media Assistance, Washington, DC.
Gillham, B. 2000, *Case study research methods*, Continuum, London.
Gillmor, D. 2005, 6 January-last update, *Tsunami and citizen journalism's first draft* [Homepage of Dan Gillmor], [Online]. Available: http://dangillmor.typepad.com/dan_gillmor_on_grassroots/2005/01/tsunami_and_cit.html [2012, 16 May].
Giridharadas, A. 2010, 13 March-last update, *Africa's gift to Silicon Valley: How to track a crisis* [Homepage of New York Times], [Online]. Available: www.nytimes.com/2010/03/14/weekinreview/14giridharadas.html [2012, 20 May].
Glocer, T. 2006, 11 October-last update, *We media speech* [Homepage of Tom Glocer], [Online]. Available: http://tomglocer.com/blogs/sample_weblog/archive/2006/10/11/98.aspx [2012, 20 May].
Greenfield, P. 2018, 22 February-last update, *Unicef deputy quits after inappropriate behaviour claims* [Homepage of The Guardian], [Online]. Available: www.theguardian.com/society/2018/feb/22/unicef-deputy-justin-forsyth-quits-inappropriate-behaviour-claims [2018, 9 May].
Hall, S.C., Jefferson, C. and Clarke, T.J. & Roberts, B. 1978, *Policing the crisis: Mugging, the state, and law and order*, Palgrave Macmillan, London.
Harding, P. 2009, *The great global switch-off: International coverage in UK public service broadcasting*, POLIS/Oxfam, London.
Hartley-Parkinson, R. 2005, 21 July-last update, *My travels: Elephants and cheeky monkeys* [Homepage of BBC], [Online]. Available: www.bbc.co.uk/lancashire/volunteering/2005/07/21/richard_in_sri_lanka.shtml [2018, 8 June].
Hayward, D. 2013, 22 April-last update, *After Boston, brainstorming better verification* [Homepage of BBC], [Online]. Available: www.bbc.co.uk/blogs/blogcollegeofjournalism/posts/After-Boston-brainstorming-better-verification-smsnyc [2013, 5 September].
Hermida, A. 2012, "Tweets and truth: Journalism as a discipline of collaborative verification", *Journalism Practice*, vol. 6, no. 5–6, pp. 659–668.
Hillier, D. & Dempsey, B. 2012, "A dangerous delay: The cost of late response to early warnings in the 2011 drought in the Horn of Africa" *Oxfam Policy and Practice: Agriculture, Food and Land,* vol. 12, no.1 pp1-34.
Hjorth, L. & Kim, K.Y. 2011, "The mourning after a case study of social media in the 3.11 earthquake disaster in Japan", *Television & New Media*, vol. 12, no. 6, pp. 552–559.
Hodson, M. 2005, "The DIY aid squad", *The Sunday Times*, 13 March edn, London.
Ibrahim, Y. 2015, "Self representation and the disaster event self-imaging, morality and immortality", *Journal of Media Practice*, vol. 16, no. 3, pp. 211–227.
International Federation of Red Cross and Red Crescent Societies, *What is a disaster?* [Homepage of International Federation of Red Cross and Red Crescent Societies], [Online]. Available: www.ifrc.org/en/what-we-do/disaster-management/about-disasters/what-is-a-disaster/ [2018, 8 June].
Jenkins, H. 2006, *Convergence culture: Where old and new media collide*, New York University Press, New York.

Johnston, L. 2015, "UGC in the newsroom: How BBC journalists' engagement with internet activists has altered newsroom practices", in *Media, margins and civic agency*, eds E. Thorsen, D. Jackson, H. Savigny, J. Alexander, Palgrave Macmillan, London, pp. 182–194.

Joye, S. 2013, "Drawing the thin line: Reflections on news media's use of user-generated content in reporting on a national disaster", in *The future of humanitarian reporting*, ed. G. Cooper, City University of London, London.

Kolbe, A.R., Hutson, R.A., Shannon, H., Trzcinski, E., Miles, B., Levitz, N., Puccio, M., James, L., Noel, J.R. & Muggah, R. 2010, "Mortality, crime and access to basic needs before and after the Haiti earthquake: A random survey of Port-au-Prince households", *Medicine, Conflict and Survival*, vol. 26, no. 4, pp. 281–297.

Lewis, P., Newburn, T., Taylor, M., Mcgillivray, C., Greenhill, A., Frayman, H. & Proctor, R. 2011, *Reading the riots: Investigating England's summer of disorder*. The London School of Economics and Political Science and The Guardian, London.

Macleod, L. 2010, 22 January-last update, *New media vital in breaking Haiti earthquake story* [Homepage of BBC], [Online]. Available: www.bbc.co.uk/worldservice/worldagenda/2010/01/100122_worldagenda_haiti_monitoring.shtml [2012, 16 May].

Manjoo, F. 2013, 19 April-last update, *Breaking news is broken* [Homepage of Slate], [Online]. Available: www.slate.com/articles/technology/technology/2013/04/boston_bombing_breaking_news_don_t_watch_cable_shut_off_twitter_you_d_be.html [2013, 7 September].

Marinovich, G. & Silva, J. 2000, *The bang club: Snapshots from a hidden war*, Basic Books, London.

Meade, A. 2017, 20 June-last update, *Please people stop taking selfies near the London tower fire disaster* [Homepage of Mashable], [Online]. Available: https://mashable.com/2017/06/20/selfies-grenfell-tower-london-fire/ [2018, 8 June].

Moeller, S.D. 1999, *Compassion fatigue: How the media sell disease, famine, war and death*, Routledge, London.

Morrow, N., Mock, N., Papiendick, A. & Kochmick, N. 2011, "Independent Evalatio of the Ushahidi Haiti project" *Development Information Systems International*. Available: https://alnap.org/system/files/content/resource/files/main/1282.pdf [2018, 19 August].

Mortensen, M. 2011, "When citizen photojournalism sets the news agenda: Neda Agha Soltan as a Web 2.0 icon of post-election unrest in Iran", *Global Media and Communication*, vol. 7, no. 1, pp. 4–16.

Naab, T.K. & Sehl, A. 2017, "Studies of user-generated content: A systematic review", *Journalism: Theory, Practice & Criticism.*, vol. 18, no. 10, pp. 1256–1273.

Newman, N. 2009, *The rise of social media and its impact on mainstream journalism*, Reuters Institute for the Study of Journalism, Oxford.

Newton, J. & Duncan, S. 2012, "Hacking into tragedy: Exploring the ethics of death reporting in the social media age", in *The phone hacking scandal: Journalism on trial*, eds. R. Keeble & J. Mair, Abramis, Bury St. Edmunds, pp. 208–219.

Nip, J.Y. 2009, "Citizen journalism in China: The case of the Wenchuan earthquake", *Citizen Journalism: Global Perspectives*, vol. 1, pp. 95–106.

O'Connor, M. 2012, 12 January-last update, *Two years later, Haitian earthquake death toll in dispute* [Homepage of Columbia Journalism Review], [Online].

Available: www.cjr.org/behind_the_news/one_year_later_haitian_earthqu.php [2016, 25 March].

Pantti, M., Wahl-Jorgensen, K. & Cottle, S. 2012, *Disasters and the media*, Book edn, Peter Lang, New York.

Paulussen, S. & Ugille, P. 2008, "User-generated content in the newsroom: Professional and organisational constraints on participatory journalism", *Westminster Papers in Communications and Culture*, vol. 5, no. 2, p. 24.

Powers, M. 2015a, "NGOs as journalistic entities: The possibilities, promises and limits of boundary crossing", in *Boundaries of Journalism: Professionalism, Practices and Participation*, eds. M. Carlson & S.C. Lewis, Routledge, London & New York, pp.186–200.

Powers, M. 2015b, "Contemporary NGO—journalist relations: Reviewing and evaluating an emergent area of research", *Sociology Compass*, vol. 9, no. 6, pp. 427–437.

Ramos, M. & Piper, P.S. 2005, May-last update, *Tsunami blogs respond to disaster* [Homepage of Info Today], [Online]. Available: www.infotoday.com/searcher/may05/ramos_piper.shtml [2018, 8 June].

Robinson, S. 2009, "'If you had been with us': Mainstream press and citizen journalists jockey for authority over the collective memory of Hurricane Katrina", *New Media and Society*, Aug, no. 11, pp. 795–814.

Robinson, S. & De Shano, C. 2011, "'Anyone can know': Citizen journalism and the interpretive community of the mainstream press", *Journalism: Theory, Practice, Criticism*, vol. 12, no. 8, pp. 963–982.

Rogers, E.M. 1978, March, *Potentials of local communications in disasters*, International Disaster Assistance—National Access to Science—National Research Conference Washington, DC.

Sambrook, R. 2005, "Citizen journalism and the BBC", *Nieman Reports*, vol. 59, no. 4, pp. 13–16.

Sambrook, R. & University of Oxford. 2010, *Are foreign correspondents redundant? The changing face of international news*, Reuters Institute for the Study of Journalism, Oxford.

Schifferes, S., Newman, N., Thurman, N., Corney, D., Goker, A. & Martin, C. 2013, *Identifying and verifying news through social media: Developing a user-centred tool for professional journalists*, Report edn, Social Sensor, London.

Seaton, J. 2005, *Carnage and the media: The making and breaking of news about violence*, 1st edn, Allen Lane, London.

Sen, A.K. 1999, "Democracy as a universal value", *Journal of Democracy*, vol. 10, no. 3, pp. 3–17.

Singer, J.B. 2005, "The political j-blogger 'normalizing'a new media form to fit old norms and practices", *Journalism*, vol. 6, no. 2, pp. 173–198.

Smith, N. 2017, 8 December-last update, *Yoga tourists seek perfect selfie as thousand flee their homes under erupting volcano* [Homepage of Telegraph Media Group], [Online]. Available: www.telegraph.co.uk/news/2017/12/08/yoga-tourists-seek-perfect-selfie-thousand-flee-homes-erupting/ [2018, 8 June].

Stephens, I. 1966, *Monsoon morning*, Ernest Benn, London.

Suran, M. & Kilgo, D.K. 2015, "Freedom from the press? How anonymous gatekeepers on Reddit covered the Boston Marathon bombing", *Journalism Studies*, pp. 1–17.

Sutcliffe, T. 2007, 2 January, *Ethics aside, citizen journalists get scoops*, 1st edn, The Independent, London.

Thompson, A. 2007, *The third wave*, Directed, A. Thompson. Distributed, Arts Alliance America. Available: https://www.imdb.com/title/tt0939669/.

Townend, J. 2012, 9 October-last update, *A journalist calls: Reporting death in a post-Leveson climate* [Homepage of Dart Center for Journalism and Trauma], [Online]. Available: http://dartcenter.org/content/journalist-calls-reporting-death-in-post-leveson-climate#.Vct86Uv_Qds [2015, 12 August].

Tuchman, G. 1972, *Making news: A study in the construction of reality*, Free Press, New York.

Tumber, H. 2014, "Back to the future? The sociology of news and journalism from black and white to the digital age", in *Media sociology, a reappraisal*, ed. S. Waisbord, Polity Press, Cambridge, pp. 63–78.

Tunstall, J. 1974, *Journalists at work: Specialist correspondents, their news organisations, news sources and competitor-colleagues*, Constable, London.

Utz, S., Schultz, F. & Glocka, S. 2013, "Crisis communication online: How medium, crisis type and emotions affected public reactions in the Fukushima Daiichi nuclear disaster", *Public Relations Review*, vol. 39, no. 1, pp. 40–46.

Vasagar, J. 2005, *Plenty of food yet the poor are starving: The two faces of Niger* [Homepage of the Guardian], [Online]. Available: https://theguardian.com/world/2005/aug/01/famine.jeevanvasagar [2018, 18 August].

Wahl-Jorgensen, K. 2015, "Resisting epistemologies of user-generated content? Cooptation, segregation and the boundaries of journalism", in *Boundaries of journalism: Professionalism, practices and participation*, eds. M. Carlson & S.C. Lewis, Routledge, London & New York, pp. 169–185.

Wahl-Jorgensen, K. & Pantti, M. 2013, "The ethics of global disaster reporting: Journalistic witnessing and the challenge to objectivity", in *Global media ethics: Problems and perspectives*, ed. S.J.A. Ward, Wiley-Blackwell, Chichester pp. 191–213.

Waisbord, S. 2011, "Can NGOs change the news?" *International Journal of Communication*, vol. 5, pp. 142–165.

Wall, I. & Chery, Y. 2011, *Ann Kite Yo Pale: Let them speak—best practice and lessons learned in communications with disaster affected communities, Haiti 2010*, Internews, London

Wall, M. & El Zahed, S. 2015, "Embedding content from Syrian citizen journalists: The rise of the collaborative news clip", *Journalism*, vol. 16, no. 2, pp. 163–180.

Wallemacq, P. 2018, *Cred crunch 50: 'Natural disasters in 2017: Lower mortality, higher cost'*, Center for Research on the Epidemiology of Disasters, Brussels, Belgium.

Wang, J. 2010, "Beyond information: The sociocultural role of the internet in the 2008 Sichuan earthquake", *The Journal of Comparative Asian Development*, vol. 9, no. 2, pp. 243–292.

Ward, S.J.A. 2005, "Philosophical foundations for global journalism ethics", *Journal of Mass Media Ethics*, vol. 20, no. 1, pp. 3–21.

Ward, S.J.A. 2008, "Global journalism ethics: Widening the conceptual base", *Global Media Journal*, vol. 1, pp. 137–137.

Wardle, C., Dubberley, S. & Brown, P. 2014, *Amateur footage: A global study of user-generated content in TV and online news output*, Tow Center for Digital Journalism, New York.

Wardle, C. & Williams, A. 2008, "ugc@BBC: Understanding its impact on contributors, non-contributors and BBC news", *BBC*. Available: www.bbc.co.uk/blogs/knowledgeexchange/cardiffone.pdf [2018, 19 August].

Wisner, B., Blaikie, P., Cannon, T. & Davis, I. 2006, *At risk: Natural hazards, people's vulnerability and disasters*, Second edn, Routledge, London.

Wright, K. 2015, "'These Grey Areas' how and why freelance work blurs INGOs and news organizations", *Journalism Studies*, pp. 1–21.

Wright, K. 2018, *Who's reporting Africa now? Non-governmental organisations, journalists and multimedia*, Peter Lang, New York.

2 Distant Suffering in a Digital World

The Background to the Changing Nature of Disaster Reporting

> Take a mega wave that has been captured on camcorder, an apocalyptic impact and aftermath plus lots of middle-class people on New Year holidays—and you have a mega story.
>
> (Liz Hunt, former executive features editor of the Daily Mail, quoted in Cooper, 2007:6)

The framing of humanitarian disasters shows how we communicate public grief in the modern world; it also demonstrates that disasters take place both in a geographical location and in a political-ideological one (Cottle, 2009). Why do some disasters become "megastories", as Hunt puts it? What influence do aid agencies in their source/gatekeeper roles have on coverage? And what part does user-generated content and social media play in mediating/mediatising disasters?

This chapter is divided into three main parts. First, it looks at the literature surrounding distant suffering—one of the most discussed areas of humanitarian reporting—to establish whether user-generated content (UGC) can alter the way we report disasters.

Second, it examines the importance of sources in covering humanitarian disasters. For many years, journalists and aid agencies had a symbiotic duopoly over disaster reporting. Journalists needed aid agencies to gain access to disaster zones. Aid agencies co-operated with journalists in order to publicly promote a cause and raise more money. This meant that a very select number of voices were heard and some forms of disaster were prioritised over others.

So, in the third section, I use Bourdieu's field theory to look at the potentially destabilising effect of the introduction of new sources to the field via user-generated content and social media, and how the established players (journalists and aid agencies) have reacted.

Distant Suffering

Why are some disasters covered and others not? And *how* are such disasters covered? These key questions are not just abstracts but can (it is theorised) make a practical difference to the amount of aid given, the foreign policy adopted and even the attitudes we have towards different people around the globe. As the former UN under-secretary general for humanitarian affairs and emergency relief, Jan Egeland, put it on the ABC programme *Nightline*, in media coverage, all disasters are not equal: "instead there is a kind of humanitarian sweepstake . . . in which . . . every night 99 per cent of them lose, and one per cent win" (Egeland, 2005 cited in Moeller, 2006:184–185).

Egeland was talking, in particular, about the Indian Ocean tsunami of 2004. But to understand distant suffering and how UGC might challenge it, there are three areas we need to look at:

1. The 'calculus of death'
2. The 'CNN effect' versus 'compassion fatigue'
3. 'Ritualising disaster coverage' and the primacy of the first-person account.

The Calculus of Death

Various studies have aimed in various ingenious ways to put together what has been termed the 'calculus of death' (Cottle, 2009:45), in order to explain what factors attract the media to cover a disaster.

Most start with Galtung and Ruge's attempt to categorise news values (1965), and refine it to explain more fully why some disasters receive the attention they do (Benthall, 1993; Allen and Seaton, 1999; Moeller, 1999; Harcup and O'Neill, 2001; Seaton, 2005; Walter, 2006). Galtung and Ruge included such factors as frequency, negativity, unexpectedness, meaningfulness, continuity, references to elite persons and nations, conflict, consonance and logistics. Harcup and O'Neill (2001) built on this nuance, with 'reference to elite nations' being subsumed under 'relevance to readers'—so that places like tourist destinations would be included (part of what made the 2004 tsunami such a massive story, for example), as well as 'elite people', including elite organisations, institutions and celebrities (2001:277–8). This, building on more recent work (Schultz, 2007; Brighton and Foy, 2007; O'Neill, 2012; Dick, 2014; Phillips, 2015), was then adapted again for a digital age to take into account 'shareability'—mainly through Facebook rather than Twitter—and including arresting audiovisuals as a new value because of this (Harcup and O'Neill, 2017).

But the recurring theme is that journalists deal with facts and numbers, and there is a myth that a certain number of people have to die for a disaster to 'count' as a disaster—the supposed tabloid approach of 'if it bleeds it leads'. This number will vary depending on *who* has died.

While such equations tend to be tongue-in-cheek[1] and are certainly not formally written down or codified by news organisations, they appear in daily practice. Almost half (49 percent) of the 290 journalists interviewed by Colombia University in 2004 said that a high death toll was the best reason to run a relief story; among non-North American journalists, the death toll angle was mentioned by 61 percent of the respondents—almost twice as often as any other possible angle (Ross, 2004).

The CARMA International report of 2006 (produced for the Red Cross's Dispatches from Disaster Zones conference) analysed six major disasters between 2003 and 2005.[2] Having looked at 2,000 articles, the report's authors came to the conclusion that "Western self-interest" was the precondition for significant coverage of a humanitarian crisis.

> The Hurricane Stanley emergency stands out as the worst indictment of the selfish Western approach to humanitarian disasters: here there is no obvious significant economic or political interest. Consequently, there is virtually no coverage of any kind . . . beyond the first few days, or coverage that focused on the humanitarian response.
>
> (CARMA, 2006:6)

Eisensee and Stromberg's 2007 study takes a rigorous econometric approach. The researchers, from Stockholm University, looked at the US government response to 5,000 disasters that occurred between 1968 and 2002, claimed 63,000 lives and affected nearly 125 million people. Like earlier studies (Adams, 1986),[3] Eisensee and Stromberg found that the number killed or affected does not necessarily influence the newsworthiness of stories. But they did find that the *type* of disaster has a direct effect on its 'newsworthiness' and thus on aid flow—with the dramatic grabbing more attention than the long-term:

> For every person killed in a volcano disaster, 40,000 people must die in a drought to reach the same probability of media coverage. Similarly, it requires forty times as many killed in an African disaster to achieve the same expected media coverage as for a disaster in Eastern Europe of similar type and magnitude.
>
> (Eisensee and Stromberg, 2007:694–695)

A disaster can be ignored because another one occurs at the same time, or occurs so close to it that it captures the media spotlight more effectively—a good example being Hurricane Stanley, which devastated Guatemala in October 2005, but which was largely overlooked in the wake of Hurricane Katrina earlier that year (CARMA, 2006).

This is what Benthall recalls as the '*phénomène de l'entonnoir*'—a 'funnel effect' where there is "only room for one overwhelming emotion a day or a week" (Benthall, 1993:28). It is rare for this to differ.[4]

As Eisensee and Stromberg note, the level of interest is not just about sheer numbers—cultural, political and geographical relevance are key too. Conflict, vested interests, secrecy and the ability to blame someone also all play their part, as well as the journalists' "imagined audience that includes 'the most important audience of all': their editor" (Kitzinger, 1999:64, also Breed, 1955).

This does not always mean that only disasters in developed countries or involving Western tourists are reported. For example, Cyclone Nargis, which hit Burma in 2008, was reported more extensively than one might expect. Here, interest was fuelled by the junta's secrecy (compared with the Chinese's relative openness about the Sichuan earthquake at the same time) and questions raised (most effectively by campaign groups such as the Burma Campaign UK) about whether the junta's actions were exacerbating people's suffering. The dramatic pictures emerging via the aid agencies[5] also whetted the newsdesks' appetites, and the few media outlets that did dispatch journalists to Burma gave large amounts of space to their reports to justify the expense of sending them there (Parker, 2008; McKenzie, 2008).

Equally, journalists may consciously or unconsciously seek patterns in events. So what at first may seem like a one-off story becomes one of a series of inter-linking stories. One of the best examples is the necrotising fasciitis outbreak in Stroud, England, in 1994, during which the bacteriologist Prof Hugh Pennington compared how the media covered one case and then another to the spread of the infectious agent itself (Pennington, 2010). While humanitarian crises rarely work exactly in that way, news editors were more likely to take interest in tsunami stories after the one in 2004. As a consequence, disasters that did not affect as many people as the tsunami, or did not have the same relevance to the West, often received minimal coverage.[6]

The 'CNN Effect' and 'Compassion Fatigue'

Yet, if the media does decide to focus on a disaster, what effect does that have? There are two opposing templates of the media's influence on emergencies: the 'CNN effect' and 'compassion fatigue'.

The 'CNN effect' (and later the 'Al Jazeera effect') is the idea that the media's coverage of humanitarian disasters can galvanise changes in foreign policy and persuade governments to intervene to alleviate human suffering, thanks to the growth of "instantaneous, real-time coverage of breaking events" (Livingston and Eachus, 1995:413). 'Compassion fatigue' argues that the media's oversimplification and the saturation of its channels with imagery of suffering makes the public increasingly indifferent to disasters.

George Kennan first linked the idea of a 'CNN effect' to the 1992 humanitarian crisis in Somalia, arguing that journalists' focus on the suffering there triggered an ill thought-out intervention (Kennan, 1993). Since then, politicians and NGOs have often suspected that control over decisions on foreign policy and aid dispersal has been lost to the media.

Livingston and Eachus summarise this view:

> The roles of the professional policy expert and diplomat have been undermined by media. To the degree that foreign policy is reactive to

> news content, the key decisions are those made by reporters, producers and editors. In this view, foreign policy decision making has become epiphenomenal to new decision making.
>
> (1995:415)

No wonder former UN secretary-general Boutros Boutros-Ghali complained that CNN was the "sixteenth member of the security council" (Minear, Scott and Weiss, 1996:4).

In the context of humanitarian disasters, the 'CNN effect' concerns not only policy and intervention, but also where aid, both from governments and the public, is directed. This explains the desire of NGOs to fund sophisticated press operations and, according to some observers, their increasing conformity to a dominant media logic, which has cost them some of their credibility and opportunities for advocacy (Fenton, 2010).

Certainly, there appears to be strong evidence that media coverage affects aid. The Red Cross's *World Disasters Report 2006* estimated that those affected by the Indian Ocean tsunami received on average $1,241 per survivor—50 times as much as the worst-funded crises during the same year (Walter, 2006).

Eisensee and Stromberg's work appears to confirm this. They concluded that the only plausible explanation was that relief decisions are driven by the intensity of media coverage:

> 2.4 extra minutes spent on the first three news segments [in a US TV news bulletin] decrease the probability that a disaster is covered in the news by four percentage points and the probability that the disaster receives relief by three percentage points. Recall that around 10 percent of all disasters are covered in the news and that 20 percent receive relief, and so the effects are sizeable.
>
> (Eisensee and Stromberg, 2007:708)

But at what stage in the proceedings did any 'CNN effect' really occur? Some argue that, in fact, the media had a bigger effect *before* 24-hour news stations with their capability to focus on live events came along. Van Belle (2008) argues that during the Cold War the media had an important impact on aid, but that around 1990 that influence declined and the so-called 'CNN effect' was illusory. This was not to argue that the media had no role but, in contrast, during the Cold War, a small amount of media coverage by major outlets could make a real difference—he estimates a single story in the prestigious *New York Times*, for example, led to an average increase in aid of US$1.76m (Van Belle, 2008:9). Afterwards, in a more diffuse media environment, only a story that was extremely salient in the news media would mean a president or prime minister would step in, as in the case of Somalia, which was erroneously used to 'prove' the media's power. Van Belle adds sardonically that, had it been the Rwanda/Burundi genocides, by contrast instead of a 'CNN effect' "we could have spent a decade arguing about a *laissez faire* effect" (Van Belle, 2008:12).

Others question at what stage the news media tends to step in. Robinson (2002) argues that the news media only got interested in the crisis in Somalia after the then US president, George H.W. Bush, had decided to send in troops. It had been aid agency and congressional lobbying rather than what appeared on TV screens that had been key to American intervention—the media acted instead an "enabler and then as a builder of support" (Robinson, 2002:62). He comes to the conclusion that a 'CNN effect' rarely occurs—and only when there is a degree of political uncertainty or 'elite dissensus', or the media coverage is framed in a highly empathetic and critical way, for example the plight of the Kurds under Saddam Hussein in 1991 (Robinson, 2002:25).

So any 'CNN effect' is most likely to occur when the reporting focuses on the sufferers. Buerk's 1984 Korem piece is a famous example—where the emphasis was on the need for outside help, rather than the political or economic reasons behind the crisis. Similarly, the coverage of the refugee crisis in Zaire in 1994 tended to avoid the complexity of the social and political issues, instead focusing on the need to get short-term relief to the refugees (Philo, 1998).

Disasters that are not framed in this way—such as the Rwandan genocide, which was largely presented as a story of ancient tribal hatred (Livingston and Eachus, 1999)—find themselves without intervention; while more empathetic coverage could have seen a bigger impact (Robinson, 2002).

The 'CNN effect' appears much more limited in its power and nuanced than may have been imagined; even the media coverage of Ethiopian famine of 1984 had more influence on the way it changed the nature of charitable giving and private philanthropy, and the policies and institutions of the voluntary sector and NGOs (Franks, 2015). And Van Belle goes on to warn that the role of the news media in disasters is frequently configured in terms of extraordinary events, such as the 2004 tsunami.

> We imagine the norm in terms of the most extreme of the rarest and most unusual events. . . . These extreme cases lead us to believe that media is an overwhelming influence on aid or that it is typical for top-level officials to be involved in the aid response.
>
> (Van Belle, 2008:2)

Yet, if the 'CNN effect' and its power to compel political leaders to act has been overstated, does that mean that 'compassion fatigue'—the idea that a glut of images of human suffering have depleted audiences' ability to care about what they see via their television screens, newspapers or the internet—has proved more durable? Is what Cottle calls the 'morally benign' media-centrism of the CNN effect giving way to the 'morally malign' media-centrism of compassion fatigue (Cottle, 2009:131)?

Compassion fatigue is the unacknowledged cause of many of the failures of journalism, according to Moeller (1999). In her view, compassion fatigue means that editors do not send reporters to cover disaster stories if they believe such stories are not of interest to their readers; and when journalists do go, their reports tend to follow a simplistic template, which results in sensationalism.

This is a heavy burden for one theory to carry—and it is unlikely that compassion fatigue is the only villain in less-than-ideal reporting. What is interesting is how it opens up a debate not only about how news organisations see disaster stories, but about how audiences respond.

Examining the Falklands War, Raymond Williams used the phrase "culture of distance" to describe how the central technical claim of television is its capacity to represent distant events and bring them close to home—but, while it was technically able to show news stories from far away, the political and cultural distances remained as great as ever. The devastation wrought by war was re-inflected by television news into an 'antiseptic' representation of reality that he defined as "childish" (Williams, 1982 cited in 2013:17).

Williams believed it was of the upmost importance to understand how television news shaped the "representation of spectacular destruction", while serving to "insulate us from reality" as we watch our TV screens at home. He talked about the "latent culture of alienation within which men and women are reduced to models, figures and the quick cry in the throat" (from *London Review of Books*, 17–30 June 1982, cited in Williams, 2013:21).

For Moeller, the distance and alienation Williams describes causes compassion fatigue. It is brought about by the simple redaction and repeated use of disaster images. She talks of the hackneyed shorthand caricatures of a crisis: the brave doctor, the ruthless tyrant, the compassionate aid worker, the vicious mercenaries, the innocent orphans and the scheming politicians.

> Each stereotype employed implies or presupposes a story line, which in turn implies or pre-supposes an appropriate political response. If the images that document a crisis are of starving orphans, the remedy is humanitarian assistance. If the images are of the brutal tyrant then the remedy is military force.
>
> (Moeller, 1999:43)

These stereotypes are used and overused; Moeller uses the phrase "overload". With victims painted in extreme word-pictures—"a dying child 'snatched' to her mother's 'bony chest', mothers 'tearing' bowls of food from starving children to feed their own"—the public's reaction is to turn off (Moeller, 1999:39).

Moeller's view, however, is criticised by Cohen (2001), who calls the idea of overload preposterous and contrary to all theories of cognition and memory. His fear is that such images become tolerable, or even normalised—when we are no longer shocked by the sight of a starving child or a corpse in a ditch.[7] For him, "the potential impact is lost because of familiarity 'we've seen it all before'. Normalisation becomes neutralisation and then indifference" (Cohen, 2001:188).

Chouliaraki also rejects compassion fatigue in favour of the term 'moral fatigue'. She writes of the "anaesthetic effect" of continually watching or hearing the repetition of a familiar soundtrack (Chouliaraki, 2006:25). In such cases, suffering becomes something to watch just like anything else, and human pain turns into "an exercise in Pavlovian compassion" (Peters, 2005:227).

Hoijer (2004) also nuances this idea by focusing on the media's frequent application of an 'us–them' perspective, in which the culture, behaviour and way of life of those affected by a disaster is portrayed as so different that they are dehumanised. Phrases like "It is a totally different culture from ours" become widespread. While Hoijer is referring specifically to coverage of the Balkans conflict, her view is equally applicable to the Western media's portrayal of African crises as tribal hatreds. In this way, our detachment is rationalised and legitimised: "Why bother about people who are primitive and uncivilized and not like us, civilized citizens in democracies?" (Hoijer, 2004:525).

Hoijer takes the word 'compassion' and defines it in four different ways in order to assess and differentiate audience response to news stories: tender-hearted, shame-filled, blame-filled and powerlessness-filled. The responses then range from empathy, to anger that such crises have been allowed to happen, to a sense of failure that one has not acted or indeed has any power to help the victims (Hoijer, 2004:522–524). By making such fine distinctions, Hoijer points out the potential limitations of Moeller's approach.

For it is clear that, despite Moeller's gloomy predictions, people still do respond to coverage of humanitarian crises—witness the millions of pounds the Disaster Emergencies Committee continue to raise year after year in appeals—albeit depending on how the media presents them. As Chouliaraki puts it:

> If this response teaches us anything, it is not that the global public exists somewhere out there ready to show its willingness to act. Rather it shows us that whom we care for is a matter of whether or not their suffering is presented as relevant and worthy of our response.
>
> (Chouliaraki, 2006:14).

The question that then remains is—who now has the power to present such suffering as "relevant and worthy of our response"? Who can overcome this distance, and involve the audience emotionally and culturally with distant 'others'? The mainstream media are no longer the only mediators of disaster zones; aid agencies can now address the audience directly, as can those caught up in disasters themselves, directly appealing to those far away through text and pictures shared via social media. Could that start to erase the distance between 'us' and 'them'?

Ritualising Disaster and the Rise of the First-Person Account

"The victim and his rescuer have become one of the totems of our age", says former head of *Medecins Sans Frontieres*, Rony Brauman (quoted in Moeller, 1999:107). "The cause of the famine is simple, the solution obvious, the victims are innocent, the rescuers heroic and those thwarting them are villainous".

The coverage of what Cohen (2001) refers to as a "famine morality play" is well rehearsed in the media. Disaster reporting is often conducted in a highly ritualised way, and "thereby become(s) infused with surrounding discourses and represented through the available cultural and media templates" (Cottle, 2009:63).

These rituals include the use of powerful iconographic images by the mainstream media, such as the identification/empathy with the dead/bereaved; the primacy of personal accounts and the related poignancy accorded to what may seem everyday items. All these elements play a huge role in "disaster marathons"—live broadcasting of traumatic events, coupled with extended coverage given by mainstream media (Liebes, 1998:73). Liebes was talking about wider notions of a disaster than the kinds of humanitarian disasters this book focuses on, but the distinction between the previous carefully staged and the unplanned event,[8] which nevertheless comes to dominate the media schedules, is apposite.

And it is incorrect to think of disaster reporting as a static genre, as recent literature has made clear (Pantti, 2011; Wahl-Jorgensen, 2013). Instead, it has taken cultural shifts and technological advances onboard.

Journalism is subject to a continual tension between "detached, dispassionate observation" while seeking to "engage feeling" (Jukes, 2017:35). While objectivity is frequently claimed as a highly prized professional value in journalism, albeit contested conceptually (Tuchman, 1972; Gans, 1980; Schiller, 1979; Schudson, 2001; Schudson and Anderson, 2009), at times of high drama such as during disasters or conflict, this may be set aside (see Bell, 1997; Morrison and Tumber, 1988).[9] This was the attitude CNN correspondent Andersen Cooper and other journalists took during Hurricane Katrina in 2005, though they came in for criticism, with the suggestion that it showed the:

> wholesale collapse of the media's own levees as they let in all the rumours, and all the innuendo, all the first-person story, because they were caught up in their own emotionalism.
>
> (Online Newshour, 29 Sep 2005, quoted in Tierney et al, 2006:74)

Yet journalists, as human beings, become overwhelmed by the situations they find themselves in, particularly with the ever-quickening news cycle. Katz and Liebes (2007) claim the marathon treatment of live events deprives journalists of the time and distance they need to think, investigate and edit. Cottle (2012), meanwhile, argues that the so-called 'beneficent embedding' of reporters in situations of strain and jeopardy (such as revolutions and political unrest, as well as humanitarian disasters) can result in them developing sympathy for, and sharing the feelings of, those they are embedded with.

Chris Cramer, CNN's international managing director (and formerly of the BBC), went further. He said he had seen an end to professional detachment in the reporting of the 2004 tsunami, that the "emotional attachment between reporter and victim has been obvious" (Cramer, 2005).

This has been emphasised by a cultural shift. Talking about the Libyan revolution and war reporting—in a way that is apposite for disaster reporting—Jon Williams, foreign news editor at the BBC, talked of a "humanity and a personalisation of the conflict"(Cooper, 2011b).[10] The shift towards human interest stories, encapsulated in the phrase "the feminisation of news", may end up favouring "more 'human' stories of civilian victims and some degree of emotional involvement, maybe allowing women reporters more space for approaching war stories in their own way and, at the same time, allowing male correspondents to respond to the intensity of the war, without the 'macho' bravado often associated with the war correspondent" (Prentoulis, Tumber and Webster, 2005:377).

Yet, reporting in a nakedly emotional way has often been criticised and continues to be associated with tabloid values by journalists themselves despite this shift (Jukes, 2017; Wahl-Jorgensen, 2018). When the BBC reporter Ben Brown awkwardly put his arm around a sobbing tsunami survivor, a furious debate erupted about whether he had crossed the line in reporting. Brown himself wrote that it had "broken the grammar of television news" [by responding], while his BBC colleague David Loyn argued, "I would make a plea for disengaged journalism. We need to keep ourselves intact" (Brown, 2005; Loyn, 2005).

In fact, despite what we might think, such shows of emotion by journalists remain comparatively rare in traditional journalism,[11]—although newer entrants such as VICE may employ it and the use of social media is encouraging it (Beckett and Deuze, 2016; Wahl-Jorgensen, 2016)—and only occurs in the most extreme circumstances. Instead, journalists tend to use ritualised forms of storytelling in order to "infuse" their work with emotion, while keeping their own feelings under control—what Wahl-Jorgensen has called the "outsourcing of emotional labour" (2013:130).

These tricks of the trade are not just deployed by tabloid journalists but widely used by the so-called 'quality' press. When Wahl-Jorgensen analysed Pulitzer-prize winning articles from 1995 to 2011, she found that the 'big' quality press were more likely to use personalised/anecdotal introductions in their prizewinning articles than the smaller newspapers (Wahl-Jorgensen, 2013).

One of the ways of infusing emotion into stories has been to 'brand' them—packaging them together under a common strapline and unifying visual design, whether in television or newspapers (Pantti, 2011).

The other change is in whom journalists see as the 'victims' of disasters; this has shifted from being those involved and their immediate families and communities towards the nation as a whole: "It has become less about 'them' and more about 'us' " (Pantti, 2011:227). The rituals seen after the death of Diana, Princess of Wales—candles, shrines, fundraising—were first apparent after the Bradford fire and have now become an integral part of disaster reporting. One could even argue that the Disasters Emergency Committee's appeal to the public after a disaster overseas is now seen as almost mandatory, for the public to be given the means to show how much we care. The BBC now regularly puts up DEC phone numbers during news bulletins as well as the free space ITV, the BBC and Sky give to a filmed appeal before bulletins.[12]

There is extensive coverage of memorials—such as the opening of a memorial to tsunami victims by the Prince of Wales and Duchess of Cornwall (Daily Mail Reporter, 2012)—increasingly including those that occur online, such as tribute videos or Facebook pages that spring up after disasters.

Dovetailing with this is the use of celebrities, not just as a way to raise funds, but as a means of articulating grief and loss. While in the past celebrities were primarily used to publicise a disaster by aid agencies, many now take to Twitter to relay their 'grief'. The *LA Times* collated a list of celebrities who had responded to the Haiti earthquake in 2010, which included hip hop star Wyclef Jean (who comes from Haiti) as well as MC Hammer, Ashton Kutcher and Demi Moore, whose connections to the island were less clear (D'Zurilla, 2010).

One of the most important rituals, however, is to profile the ordinary person—the heroes and heroines among the rescue workers and survivors, the tragic tales of those who have suffered loss.

This has long been a tradition of journalism although journalists' focus on the personal when interviewing survivors or the bereaved has intensified in recent years (Wardle, 2006; Coward, 2013). But this trend has particularly been facilitated by the use of user-generated content—in particular, raw footage that appears unmediated or so-called 'real' voices. During the 2004 tsunami, much of the coverage's power came from camcorder footage—it was a turning point (Gillmor, 2005). The BBC received 400,000 messages on its message boards within just ten days of the tsunami (Cooper, 2007); by the time the Haiti earthquake hit five years later, new media platforms such as Twitter, Facebook and Skype were crucial in delivering early information about damage and relief efforts (Brainard, 2010). Many websites, such as the *Guardian*'s, ran liveblogs and Twitter feeds, allowing those on the ground to tell their own story (Bunz, 2010). By 2015 the Nepal earthquake was being covered via the livestreaming app, Periscope (Cooper, 2015).

In an electronic age dominated by visual rather than written communication, images have a visceral public impact that no other medium can make: "if only they knew" replaced by "now they can see" (Cohen, 2001:185). Images can be replayed endlessly on a loop, generating "the mediatised *enabling* of crisis-emotions among those who were not physically present at the event" (Frosh and Pinchevski, 2009:301). Also, there is the belief that UGC gets nearer to the sight, sounds, smell and touch of a disaster, in a way that traditional journalism cannot. Of war journalism, Taylor said:

> Real war is about the sounds, sight, smell, touch and taste of the nasty, brutal business of people killing other people. Media war is literally a mediated representation of that reality but which is ultimately a third party or audio-visual 'snapshot' of it.
>
> (Taylor, 2003:69)

Compare that to Fredo Dupoux's gabbled—and garbled—tweet "Oh shiet [sic] heavy earthquake right now! In Haiti!" (Macleod, 2010). From the shaky footage of the 2004 tsunami wave captured on camcorder to the livestreaming via Periscope of Hurricane Matthew in 2016 (Burrell, 2005; King, 2016)—the very unprofessionalism of the video and words appears to sidestep the mediated distance of traditional journalism.

It also challenges the idea of the 'insider' journalist. For years, reporters have operated in a community that creates the cultural authority to tell society its news (Zelizer, 1993). Journalists constantly negotiate and agree on the 'right' story, while

> the culture of journalism functions as an autopoietic or 'self-organizing' social system. Newcomers are primarily expected to adapt themselves, and to adopt the dominant (ideological) perception of what journalism is.
>
> (Deuze, 2007:169–170)

As quoted earlier, the new media world then becomes a paradigm shift, "in which once the media was the center of the universe and now the user is the center of the universe" (Robinson and De Shano, 2011:977).

Robinson and De Shano somewhat overstate the case, but supporters of user-generated content see it as a means of escaping some of the mediative processes; a move away from Chouliaraki's "anaesthesia" of traditional reporting. Audience focus groups questioned about UGC consistently respond positively, seeing it as more authentic, real and emotional. Journalists also seem to value the perceived aesthetic value of the typically grainy footage of UGC, signifying access to the real (Wahl-Jorgensen, Williams and Wardle, 2010; Williams, Wardle and Wahl-Jorgensen, 2011).

Finally, supporters of user-generated content potentially allow a greater diversity of stories to be heard; Cohen raises the idea of the 'active bystander' in journalism, when describing the distribution of video cameras to human rights observers.

> A row of silent witnesses . . . watching wrongdoing often putting themselves at risk is a powerful image. They are active bystanders—powerless to intervene but a reminder to perpetrators that not everyone approves or colludes and their future denials will be countered by another testimony.
>
> (Cohen, 2001:256)

And Curran notes that:

> the internet makes it easier to transmit alternative views to a wider audience . . . alternative voices tend not to be heard in the main square of the electronic public sphere but they still get heard in some online back streets.
>
> (Curran, 2010:286)

However, some of the claims for UGC's influence on the coverage of humanitarian disasters have been over-enthusiastic.

First, a digital divide still exists; access to this technology—and the skills to use it—are not democratically distributed (Andén-Papadopoulos and Pantti, 2011). A study on BBC convergent news of the Haiti earthquake shows that while its web-stream invited the contributions of those affected as the main source of news, only eight web-stream entries came from 'average people' in Haiti (and it is unclear how many of these were actually Haitians). The remaining 42 were attributed either to Western NGOs or to Westerners indirectly touched by the earthquake (Chouliaraki, 2010:15).

This personalisation also predates the use of UGC (Hoijer, 2004). News media was already focusing more on civilians as victims of conflicts than before. According to the former BBC war reporter Martin Bell, the reporting "changed fundamentally" from focusing on the military aspects of war, such as strategies and weapon systems, to putting more emphasis on people—"the people who provoke them, the people who fight them and the people who suffer from them" (1998:15–16).

And this 'personalisation'—one of UGC's great strengths—does not always have the desired effect; "the aim to break down the 'us and them' dichotomies—the structural interests of 'people like us'—are counterpoised against the suffering of strangers" (Allan, 2004:361). Chouliaraki highlights the fact that we live in a society where

> our own private feelings are the measure against which we perceive and evaluate the world and others. . . . While news becomes part of this 'culture of intimacy' it implicitly allows us to focus on our own sufferings and disregard those 'others' outside our own horizon of care.
>
> (Chouliaraki, 2006:13)

There is simply a limit to the empathy we can have. For example, close-ups of children suffering hunger may seem to establish some proximity between child and audience, but focusing on these children's emaciation " 'fetishizes' their fragile bodies and maintains a radical distance between them and the spectators" (Chouliaraki, 2006:87). The UGC creator who films those suffering in a flood, an earthquake or a tsunami may inspire empathy momentarily, but we cannot truly escape the idea that this is "distant suffering"; it could be us—but it is still not.

Being part of the action can also limit the story told. Gralnick says that reporters embedded during a conflict "have no idea how the war is going, only how [their] war is going" (Gralnick, 2003). As Seaton notes, citizen journalists' work can be "simultaneously exact representations of what they could see and partial because limited by their inability to take into account anything of the wider situation beyond their own harrowing experience" (2005:xxiii).

So, UGC may show survivors or victims of disasters—people living without aid—but may not provide an understanding of why aid agencies or governments have failed to act (Curran and Seaton, 2010:322). It cannot, therefore, be seen as a total panacea for the failures of disaster reporting in the past.

The Role of the Source in Reporting Disasters

This is a summarised version of the necessary ingredients for covering a famine, listed by George Alagiah (then developing world correspondent for the BBC) at a Dispatches from Disaster Zone conference in 1998:

> Starving child (preferably crying)
> Feeding centre (complete with mothers with shrunken breasts)
> Aid worker (usually white, usually a woman, battling against the odds)
> Reporter (breathless and shocked, saying how awful it is).
> (Alagiah, 1998, adapted by Cooper, 2007)

What Alagiah calls 'template reporting' is familiar to anyone who has watched television news bulletin about a food crisis. But the recurring question, as already discussed, is *why* disasters such as food crises are framed in these ways. A crucial related question is how the sources that journalists use affect how a story is framed.

In a food crisis, for example, the choice of sources may seem obvious: an aid worker is usually interviewed because they are on the ground, at the centre of the action and likely to speak English; and those at the feeding centre are asked the classic "How do you feel?" experiential-type questions, as a quick way to ensure the viewer empathises with people who have travelled miles, desperate for food.

But the use of sources not only helps select which stories—whether disaster, environmental, political or health—achieve a high media profile, it also plays a huge part in framing the story—sometimes in ways that are less than ideal (Manning, 2001). For example, an aid agency has an interest in making the disaster sound as bad as possible to motivate people to send aid, and—if you are very cynical—keep them in business (Birrell, 2012).

The corporate culture—the 'news factory' within newsrooms themselves—also plays a crucial role story selection (Manning, 2001). This could involve the sub-editor acting as a 'gatekeeper', choosing certain stories and spiking others, which means that many stories never see the light of day (White, 1950).

The literature around sources still owes much to the work of Hall et al (1978). They argued that the pressure that journalists found themselves under to meet deadlines, and the professional demands of objectivity/impartiality, meant those in privileged institutional positions—so-called 'primary definers'—were often used as sources most frequently. Or, as Gans put it: "The president of the United States has instantaneous access to all news media whenever he wants it; the powerless must resort to civil disobediences to obtain it" (1980:119).

Hall et al saw journalists as accepting and framing stories in a way that did not critique those in power, but backed up their version of events. However, this perspective has been challenged. Schlesinger (1990), for example, noted that Hall did not take into account the fact that the so-called 'primary definers' changed over the years. Anderson (1993, 1997) also pointed out that, in the wake of health and environmental scares, outsiders—pressure groups such as Greenpeace—had an opportunity to set the agenda. In the case of disasters, this would mean international aid agencies, such as Save the Children and Oxfam.

Journalists constantly switch between being controllers and victims of time. It leaves them at the mercy of the story breaking late and prioritises the 'helpful source' who has the resources to staff press offices 24 hours a day (Schlesinger, 1978). Certainly, 'helpful sources' can include Anderson's 'outsiders'; aid agencies have invested in their media teams to the extent that, by 2007, Oxfam had more press officers than the *Independent* newspaper had general news reporters (Cooper, 2007).

This is not to say that the official line is always accepted. Consider, for example, the MMR crisis, when one (now retracted) paper in the *Lancet* questioned the vaccine's safety (Wakefield et al, 1998); despite the efforts of the health establishment and the government, newspapers continued to promote Dr Andrew Wakefield's theories.[13]

Manning puts it this way:

> The institutional arrangements of corporate capitalism and the state allocate many of the best cards but not all the cards to the powerful, in

> the struggle around news agendas. The less powerful and the politically marginal do get a look in, but how and to what extent are empirical questions.
>
> (2001:ix)

Schlesinger and Tumber (1994) also claim Hall overstates the "passivity of the media as recipients of information from news sources" (1994:19) and that Hall's work showed insufficient curiosity about the processes that the sources themselves go through to get their story out.

Even with a sudden-onset disaster, any pressure group or campaigning organisation undergoes an (often pretty brutal) inner debate about what information they are going to reveal and when. While it would be wrong to suggest that aid agencies can fully determine the media agenda, they have become increasingly professional in order to influence the media to prioritise certain stories. As a senior press officer at Oxfam commented: "We do not believe aid changes the world—politics does" (Cooper, 2011a:42).

Aid agencies now employ increasingly sophisticated communication techniques and invest heavily in media professionals. They deploy celebrities to promote causes and describe themselves in terms of their 'brand' (Cottle and Nolan, 2009). This is with the purpose of raising money and the public's interest in macro issues of poverty and inequality.

So no discussion about disaster reporting can be complete without looking at the perspective and actions of journalists' sources. The media strategies of aid agencies and other NGOs are worked out in what Schlesinger and Tumber call a "quite specific, if implicit, conception of the public sphere" (1994:104). For acute appeals, the terrestrial television channels (BBC One, ITV1 and Channel 4), elite radio programmes (such as BBC Radio 4's *Today*) and satellite channels (Sky News, the BBC News Channel, Al Jazeera English) are seen as the key to raising awareness.

Midmarket and tabloid papers, and women's magazines, are targeted for fundraising articles and the broadsheets for campaigns, as the popular press is still seen as resistant to discourse about policy. But, for many, social media is increasingly important.

If there was any debate about this, the Haitian earthquake proved how important an effective social media presence could be. As mentioned earlier, according to the Twitter-tracking service Sysomos, between 12 and 14 January, 2.3 million tweets included the words 'Haiti' or 'Red Cross'. There were also 189,024 tweets that included '90999', the number that could be used to text message a donation to the Red Cross (Evans, 2010). The result is that whole media strategies now revolve around social media, as will be discussed later on.

Gans (1980) also argues that affluent organisations (and, as can be seen, many aid agencies *are* affluent)[14] do 'preschedule' their activities to satisfy news organisations' desire for planned stories, or can supply spokespeople at short notice. Aid agencies may also be ingenious enough to create so-called 'media events'—activities that exist solely or mainly to be covered by the media.[15] Boorstin (1992) has called such stunts 'pseudo-events', which should not be treated as news. For smaller organisations, however, such 'politics of spectacle' may be the only way they can get noticed.

The apogee of NGO stunts came with Make Poverty History in 2005 and the 2013 Enough Food for Everyone IF campaign (often referred to as Make Poverty History 2). These global campaigns saw the most prominent UK aid agencies come together to raise awareness about trade justice, poverty and overseas debt, using a combination of grassroots stunts and sophisticated marketing and advertising appeals to ensure these issues became part of the media agenda. Manning points out that, in recent times, the symbolism that such stunts deliver "may be as important as the rigour of an intellectual argument" (2001:10); this "politics of spectacle" promotes the idea that NGOs are 'outsiders', even though they are much closer to power than the public may realise.

The potential problem, according to Cottle and Nolan (2009), is that aid agencies are becoming "embroiled in a media logic that is far from the aims and ideals of humanitarianism". Fenton (2010) warns that by conforming to the dominant media logic, agencies are actually losing opportunities for advocacy and any credibility they possess by their distinction from the elite.

The way aid agencies provide information now goes well beyond press releases, stunts or providing commentary on a 24-hour basis. They have increasingly sophisticated relationships with the media, a development which can be traced to the late 1990s. Agencies had seen their reputations tarnished by how they had handled the press in places like Goma (Gowing, 1998). They had also come to realise that it was an increasingly competitive market place, and hired journalists to professionalise their press offices pushing the idea of press officers as 'fireman' reporters, not only facilitating media requests but producing written and visual content themselves (Cooper, 2007). So those aspiring to be primary definers were acting in a very different way, adopting the persona and activities of reporters. Facilitated by increasingly sophisticated and affordable mobile technology, this change occurred almost in parallel with the rise of the creation and publication of content by ordinary people.

Applying Field Theory

Few would disagree with the fact digital technologies are altering the way we can get access to news and create different dynamics in the way news is produced and published (Tumber, 2014).

The formal study of how news organisations "produce news products" (Schudson, 1989:264) dates back to the early 1950s with the concept of 'gatekeeping' applied to journalism (White, 1950; Gieber, 1964), which laid the groundwork of theory around news as a construction. Limitations of the gatekeeping approach however led to further application of theories which focuses on how journalists are constrained by "organisational and occupational routines" (Schudson, 1989:266).

This was forefronted in sociological studies of the 1970s and 1980s in particular Tunstall (1974) which focused on the organisation and routine management of news and was thus able to outline general practices, journalists' behaviour and the macro-level structure of news production (Tumber, 2014).

But the methods and production of journalism are changing in a world of new media technologies. How can we best explain and understand media concentration, personalisation of the media and the changes wrought by the fact that today anyone can self-broadcast/publish? What effect does user-generated content and social media have on the boundaries of journalism? Just as Tuchman and Tunstall described the importance of journalists' routines, practices and power relationships in the past, how do we accommodate the new approaches to journalism in the early 21st century?

These routines and practices have undoubtedly been affected by the increasing financial uncertainty. Those who take a political economy approach see the "interplay between economic organisation and political, social and cultural life" (Golding and Murdock, 1991:73) in which the balance of power between capitalist enterprise and public intervention has become particularly acute. Recent scholars have noted that the media industry has for some time been under intense pressure, with a 'perfect storm' (Barnett, 2009) having been created by a decline in advertising revenue in recent recessions, the migration of advertising revenue to the internet and the fragmentation of audiences. The result has been a continued concentration of power in the larger, more established news providers, despite the hopes that the internet would bring more diversity (Fenton, 2010) while for many journalistic organisations there has been an increased dependence on news agency copy and PR sources (Davies, 2008:54, 74).

While many studies have been able to demonstrate that news coincides with, and reinforces, the political and economic elite (Murdock, 1973) this tends to explain the big picture, rather than the day-to-day practices of journalists. So sociological researchers built on both the Marxist approach, championed by those such as Herman and Chomsky (1988) or the Gramscian idea of hegemony but continued to examine how journalists actually operate on the ground—their patterns of behaviour and autonomy.

The integration of digital technology has caused sharp divides in critical thinking however. McChesney divided scholarly and public opinions on the role of the Internet into two sides in contemporary society. On one side were the "celebrants"—who argued that digital technologies promote global democracy and innovation. This is contradicted by the digital "sceptics", who "fear deeply that the Internet is re-creating people in technology's image, flattening our intelligence and lessening our creativity" (2013:12). McChesney, however, suggests that both sides are limited as neither engages sufficiently with issues of political economy (Broad, 2013).

This has been exacerbated by the dominance of a few online brands such as Facebook, Google and Twitter which are not traditional 'news' organisations but who have taken over the role of disseminating 'news' to many. This has resulted in a continuing centralisation of media—by November 2016, the Pew Research Center was reporting that 79 percent of online Americans now had a Facebook account (Greenwood, Perrin and Duggan, 2016). Those early hopes that many more diverse voices would be heard appear dashed—as Benkler presciently puts it a decade ago: "the Internet diverges from the mass media much less than we thought in the 1990s and significantly less than we might hope"(2006:214).

In trying to understand the upheaval that journalism is going through and the new structures and practices used for producing news, Habermas's public sphere and Bourdieu's field theory may prove a useful theoretical framework. While both theories owe a debt to the Marxist concept that the ideas of the age are those of the ruling classes, they provide more flexibility and insight into the economic, cultural and technological pressures that are all brought to bear on the way news is reported, despite Bourdieu's primary interest in restricted rather than large-scale production (Hesmondhalgh, 2006).

They offer the possibility that new agents may have an influence, and take into account changes in technology. And, while recognising there are different and competing sources, these theories still accept that there is not necessarily equality of access and resources for all sources.

Habermas's (1989) concept of the 'public sphere' provides a rather gloomy view of public engagement today. From a high point in the 17th and 18th centuries, when Habermas believed there was a possibility of rational communication involving the exchange of ideas after people had sprung free from the feudal control of church and state (thanks to the emergence of coffee shops, newspapers, journals, debating societies, libraries and capitalist markets), he argued that, by the 19th century, capitalist societies were contributing to a re-feudalisation, with advertising, PR and commercial sponsorship replacing monarchs, church and feudal lords (Manning, 2001). Today social media could be seen to form part of this sphere.

Of course, many have criticised the idea of a 'rational' public sphere—for example, Manning asks if a radio phone-in show has to live up to Enlightenment standards of reasoning. Others point out that the coffee shops of the 18th century excluded people on grounds of class and gender (Fraser, 1990; Eley, 1992; McLaughlin, 1993). Curran and Seaton (2010) and highlight the emergence of 'counter public spheres' such as the radical working-class press of 19th century. Meanwhile, Gitlin asks whether we can even speak of a 'public sphere' in an increasingly diverse, complex series of societies. He suggests that what has emerged instead are numerous public 'sphericules' (Gitlin, 2002:173).

But some argued that a Habermasian public sphere could emerge, thanks to the internet. Early researchers saw the worldwide web as presenting a huge opportunity, with the sheer weight of information making news journalism more vital than ever (Bardoel, 1996). Others optimistically thought the growth of online data would transform the relationship between journalists and their sources, because the journalists would have as much access to information as the public officials they were writing about (Koch, 1991).

Meanwhile, Livingstone and Lunt (1994) argued that broadcast audiences were becoming more confident and skilled at using tools such as email and the internet to challenge the public accounts of political elites on radio and television.

But often it is individuals rather than social groups or organisations who have more of a voice; thus they tend to have less power, and are often talking in a personal capacity about their own experience as opposed to having a more analytic role (Manning, 2001). Cottle (2000) agrees; he argues that, while "ordinary voices" often find their way into TV news items, they are rarely granted an opportunity to develop their arguments or points of view at length—or challenge political or expert authorities. Instead, they become "the voices of the side effects" (Beck, 1992; Cottle, 2000:30)—to symbolise the human face of a news story.

So, while ordinary voices may appear prominent, they are most often heard in short news pieces and hardly at all in more discursive reports (the live studio interview, for example). Again, those who do speak are speaking about their own experience, as opposed to giving analytic accounts.

Yet, Bourdieu's field theory suggests that there can be a potential shift in the journalistic field—the realm where he says actors struggle for autonomy—as and when new agents gain access.

Writing in *On Television*, Bourdieu defined the journalistic field as

> a structured social space, a field of forces, a force field. It contains people who dominate and people who are dominated. Constant permanent relationships of inequality operate inside this space, which at the same time becomes a space in which various actors struggle for the transformation or preservation of the field. All the individuals in this universe bring to the competition all the (relative) power at their disposal. It is this power that defines their position in the field and as a result their strategies.
>
> (Bourdieu, 1998:40–41)

Field theory is particularly relevant because of its interest in practices of journalism. Bourdieu argues that journalists and the organisations they work for can be seen as being at various points between two poles, relating to economic and cultural capital. Those positioned closer to the economic capital pole tend to be wealthier, but also conservative because they have less autonomy from the field of power. By contrast, those who have more cultural capital are often seen as possessing less wealth, but having more autonomy (Bourdieu, 1998).

It has been argued that the mainstream media in the UK does not fit fully into this model, mainly because of the BBC, which is both wealthy and conservative (Berry, 2013; Wahl-Jorgensen et al, 2013; Ayton and Tumber, 2001). The BBC also possesses immense cultural capital because of the quality of its journalism, and because of its privileged position as the national broadcaster (Wright, 2014).

For me, the most salient question is: given that Bourdieu was mainly writing in a pre-internet, pre-social media age, can the questions he raised be applied in such a rapidly fragmenting media environment?

The internet is, of course, only the field's latest entrant; television and radio also created audiences and spaces for new forms of journalistic practices. None of these technologies were initially a journalistic medium, but became incorporated into the field and transformed both the field and practices with in it (Krause, 2011:100; Barnard, 2012). Thus, a field approach can invite us to "think about media and cultural production as a *modus operandi* and not only as an end product (*opus operatum*)" (Neveu, 2005:203).

For Bourdieu, boundaries are often fuzzy and contested, and this is where change occurs; boundary (re)negotiation is a key struggle in the journalistic field at the moment, involving new technologies and those with the power to best use them (Carlson and Lewis, 2015). Sterne understands this as technologies acting effectively as 'subsets' of the habitus or "organised forms of movement" (Sterne, 2003:370).

Technologies may 'sediment' social relationships—for example, as Sterne puts it, radio is something that traditionally ordinary people listened to rather than created, but the internet and social media allows people to broadcast/publish themselves, which may constitute the emergence of whole new fields, as well as disruption to existing ones. For example, Arrigara and Levina (2008) define an online cultural field as a

> group of individuals who use a common IT platform, share a cultural affinity and who by engaging in the production, dissemination and valuation of user-generated content on this platform have an influence on the trajectory of other individuals in that field.
>
> (2008:2)

It is easy to get bogged down, however, in the details of these new technologies and forget what they represent. Bourdieu saw technologies as fundamentally social, believing that what generates technology is social practice (Davison, 2004:87; Myles, 2010). So, the use of any particular communications technology—whether a mobile phone or social media network—will change due to social variations, but this must be understood in the context of the way technologies are implicated in the 'action' of all fields.

Early research, for example, looked at mobile phone ownership as a way of gaining social capital (Henderson et al, 2002) but, as Fortunati presciently argued, the idea of a mobile as status symbol quickly became outdated[16] and it soon became seen as a "citizenship commodity" (2002:523). Mobile communication acted as "privatis[ing] public space", but it can also be used to participate in a public field—what has been called 'public flocking' behaviour (Rheingold, 2007:13).

So, while technologies do not change societies or social processes just by coming into being, they can affect social processes through both mundane and innovative uses (Earl and Kimport, 2011). Thanks to the ubiquity of the mobile phone and social media it seems "our lived reality cannot be experienced separate from or outside of media" (Deuze, 2011:140). And while the journalistic field has a significant impact on others, those who use social media and create user-generated content may also have an impact on journalism. This may be resisted and remoulded—as Singer (2005) found in her study of j-bloggers, who aimed to 'normalise' blogs to traditional journalistic practice—but the possibility of change is there (Hermida, 2009; Lasorsa et al, 2012).

Who are these UGC creators? Several studies that looked specifically at news comments have mapped and analysed users, finding them to be more educated and often older than internet users in general (Almgren and Olsson, 2015; Bergstrom, 2008; Freund, 2011; Nagar, 2011). To understand the extent to which users participate and how they participate, it is crucial to consider the extent to which they are actually invited to do so.

For, just as with print, television and radio, when users participate online, this does not happen in a neutral environment isolated from outside influences (Almgren and Olsson, 2015). In fact, participatory practices—such as commenting on liveblogs, joining a Facebook discussion or contributing to a hashtag on Twitter—generally take place in spaces that have been carefully structured. Even if they appear to be uncontrolled by existing elites, they are often swiftly commandeered by media elites, who want to control and monitor these contested boundaries.

Freedman argues that the internet has not democratised what appears in the media, but merely perpetuated a "commodity exchange" controlled by existing elites (2012:88); Fuchs nuances this by pointing out that most online spaces are offered to users by corporate social media.

> If there are for example asymmetries in terms of visibility and attention, then it is questionable to argue that corporate social media are truly participatory. It is therefore not enough to stress enabling and limiting potentials of the Internet, but one rather needs to analyse the actual distribution of advantages and disadvantages.
>
> (2013:2)

Yet, it would be overly pessimistic to assume this means no real change. Bourdieu saw the journalistic field as a constant struggle, and was concerned with how agents experience change when there is a disjunct between the *habitus*[17]—understanding or 'having a feel for' the journalistic game (Bourdieu, 1990:66)—and current conditions in the field.

Field theory argues that journalistic fields do not always reinforce the power *status quo*. Under some conditions, they can transform power relationships in other fields (Benson and Neveu, 2005:9).[18] Klinenberg argues that field theory is particularly useful for identifying the relationships and forces that establish the positions of media producers and consumers. It can therefore help to map the range of challenges faced by activists and other groups (such as aid agencies) interested in effecting change (Klinenberg, 2005:174).

Russell's (2007) analysis of the 2005 riots in Clichy-sous-Bois appeared to show a destabilisation in the field; the citizen journalists and bloggers transforming the coverage of the riots and challenging the French media's self-censorship (2007:293).

However, Vos et al (2012), in their critique of political bloggers, were less optimistic about any such transformation. They found that bloggers tended to accept the received journalistic 'doxa' (the so-called nature of the field) and thus favoured stability in it. Such bloggers were very keen to distinguish themselves from the mainstream media—often by using pejorative terms for it, such as 'deadwood press'—and saw themselves as an autonomous force. But by criticising traditional journalism they were effectively holding it up as an authority. So Vos believes that the relationship between new and old media has become one of "rivalry not reciprocity" (2012:4), with bloggers acting as self-appointed "watchdogs of the watchdogs" (Hayes et al, 2007).

Russell does not totally disagree with this conclusion—she sees a blurring of the lines between the two. But it is not just that bloggers and creators of UGC have conformed to mainstream media—there is two-way traffic. As she puts it:

> Digital DIY news products and practices influenced and were influenced by mainstream media: new and old-media journalists adopted one another's norms, practices and language in attempts to gain or maintain legitimacy.
>
> (Russell, 2007:296)

What is clear, however, is that we begin to see the suggestion of previously separate groups—journalists, their traditional sources (aid agencies) and their 'audiences'—starting to adopt different roles; this is not merely ordinary people being able to tell their own stories to a wider audience than would have been possible before the advent of easily accessible mobile technology.

It concerns aid agencies disrupting the field by becoming publishers and retelling the story, or absorbing potentially troublesome material from survivors. It also involves journalists, who have faced credibility problems in recent times,[19] trying to appropriate social capital by adopting the tropes associated with UGC creators, who are traditionally seen as more 'authentic' and truth telling (Wahl-Jorgensen, Williams and Wardle, 2010; Williams, Wahl-Jorgensen and Wardle, 2011).

What does this mean in practical terms? Do these disruptions mean that the problems laid out in the first section of this chapter—the ritualisation of disaster reporting and the subsequent distancing this can produce—are lessened or potentially made greater? The next chapter uses empirical data to see how journalists and aid agencies currently see this situation.

Notes

1. For example, that for the British media, one UK death in the same street equals ten in the next town equal 100 deaths in France and 1,000 in India (Franks, 2006).
2. These were: the earthquake in Pakistani Kashmir (8 October 2005), Hurricane Stanley (1 October 2005), Hurricane Katrina (23 August 2005), the Indian Ocean tsunami (26 December 2004), the earthquake in Bam, Iran (26 December 2003), and the humanitarian crisis in Darfur, Sudan (since February 2003).
3. Adams measured the amount of coverage that disasters received in the American media compared with the miles that they were distant from New York City.
4. What was extraordinary about the coverage of the 2008 Sichuan earthquake and Cyclone Nargis was that for once two disasters of similar severity in faraway places were covered extensively side-by-side in the British media (Cooper, 2008).
5. For a fuller description of this, see Chapter Three and discussion of how the BBC used footage from Jonathan Pearce of Merlin to cover Cyclone Nargis.
6. Interviewee 38, an international editor for a commercial broadcaster, recalled how, after the 2004 tsunami, news editors were on permanent alert for the next one: "Three months after the tsunami, there was another earthquake and tsunami alert in Indonesia. We were sent very quickly. There were loads of us at the airport—I remember seeing Ben Brown [another star reporter] from the BBC. We all got on a plane, we went to Singapore. But by the time we'd landed and called in, it was quite clear that this was not a Boxing Day tsunami-type event. And in fact we spent about 24 hours there, realised it would cost us an enormous amount of money to cover this properly. And on the basis of 'it's only a thousand dead', I'm afraid we all pulled back."
7. See also Cohen and Seu's pilot study at Brunel University, where focus groups were shown Amnesty International appeals; those in the focus groups recognised an accumulation of atrocity (Cohen and Seu, 2002).
8. I would argue however that NGO media teams now aim to facilitate disaster coverage which seems like a "disaster marathon" when in fact it has more in common with a 'media event'.
9. This is perhaps separate from BBC reporter Martin Bell's 'journalism of attachment', which employs an overt normative approach for war reporting. Bell wrote that after his experiences in warzones he could no longer endure what he termed the BBC's dispassionate 'bystander journalism' and instead argued in favour of a 'journalism of attachment' (1998:15–16). He differentiated this from campaigning journalism, but framed it in moral terms, calling it a journalism "that will not stand neutrally between good and evil, right and wrong, victim and oppressor" (1998:16). This normative debate is often obscured in 'natural' disasters.
10. It's also worth noting as the Channel 4 veteran reporter Lindsey Hilsum points out that it is easier for male reporters to articulate emotion. She says: "I think you'd be hard-pressed to find a consistent distinction between men's and women's

reporting of wars and revolutions. But I would say that when a man does the weepy, human side he is regarded as empathetic and sensitive, but a woman may be perceived as 'not coping' if she shows emotion. So women broadcasters have to be very careful not to play into people's stereotypes" (Cooper, 2011b).

11. It is not completely uncommon. Andersen Cooper has made his name through what we might call 'involved journalism'—witness his rescue of a small child during the Haiti earthquake (Cooper, 2011a).
12. The decision of the BBC not to air a DEC report for Gaza in 2009 on grounds of editorial impartiality provoked fierce debate, in part because DEC appeals had become so ubiquitous. It's worth remembering, however, that this was not the first time the BBC had turned down an appeal—it did the same in Lebanon in 2006 (Tryhorn, 2006).
13. The *Daily Mail*'s parliamentary sketch writer, Quentin Letts, had a child with autism, whom Letts said had developed their condition after receiving the MMR vaccine (Davis, 2004). The paper campaigned on the MMR issue, particularly focusing on the then prime minister Tony Blair's refusal to say whether his son Leo had been vaccinated (White, 2002).
14. Save the Children UK, for example, grew from a £140m turnover to £300m turnover in the space of five years according to its chief executive Justin Forsyth (interview, London, 17 January 2012).
15. For example, in 2007 Oxfam spent a great deal of time in internal meetings debating whether it should run an appeal for Darfur/Chad; it spent six weeks planning its appeal. It clearly defined its target audience—wealthy professionals, predominately females, who read the liberal press and watched the Ten O'Clock News. Its 'stunt' involved paying the celebrated war photographer Don McCullin to photograph refugees in Darfur and Chad, selling his story as an exclusive to the BBC, while also facilitating a trip for the *Guardian*. To maximise media coverage it also asked the advertising agency Saatchi & Saatchi for some *pro bono* work. The ad agency came up with the idea of a replica refugee tent built in London to act as a live TV backdrop.[15] The resulting media coverage included an exclusive report on the BBC's *Ten O'Clock News*, a photo essay in the *Independent* and coverage from all rolling news stations (Cooper, 2007).
16. Ninety-three percent of 16- to 24-year-olds in the UK now use a smartphone (Ofcom, 2017).
17. Bourdieu defines 'habitus' as a property of actors (whether individuals, groups or institutions), which comprise a structured and structuring structure (1984: 167–71). The 'structured' refers to past experiences, the 'structuring' that habitus helps shape one's present and future practices and it is 'a structure' in that it is systematically ordered (Maton, 2008:50). A journalist's practice results from the relation between one's disposition (habitus) and one's position in the field (capital) with the current state of play in the field. Thus practice is not simply "the result of one's habitus but rather of relations between one's habitus and one's current circumstances" (Maton, 2008:50–51).
18. Bourdieu uses the metaphor of physics to try to explain disruption to the field: "The more energy a body has the more it distorts the space around it and a very powerful agent within a field can distort the whole space, cause the whole space to be organized in relation to itself" (Bourdieu, 2005: 43).
19. This is particularly pertinent in the UK after the revelations of the phone hacking scandal and the subsequent Leveson inquiry.

References

Adams, W.C. 1986, "Whose lives count? TV coverage of natural disasters", *Journal of Communication*, vol. 36, no. 2, pp. 113–122.

Alagiah, G. 1998, 27–28 May, *Dispatches from disaster zones: Reporting humanitarian disasters*, conference proceedings, British Red Cross, London.
Allan, S. & Zelizer, B. 2004, "The culture of distance: Online reporting of the Iraq war", in *Reporting war: Journalism in wartime*, eds. S. Allan & B. Zelizer, Routledge, London, pp. 374–366.
Allen, T. & Seaton, J. 1999, *The media of conflict: War reporting and representations of ethnic violence*, Zed, London.
Almgren, S.M. & Olsson, T. 2015, "'Let's Get Them Involved' . . . to some extent: Analyzing online news participation", *Social Media Society*, vol. 1, no. 2, DOI:2056305115621934.
Andén-Papadopoulos, K. & Pantti, M. (eds). 2011, *Amateur images and global news*, Intellect Books, Bristol, UK.
Anderson, A. 1993, "Source-media relations: The production of the environmental agenda", in *The mass media and environmental issues*, ed. A. Hansen, Leicester University Press, Leicester, pp. 51–68.
Anderson, A. 1997, *Media, culture and the environment*, UCL Press, London.
Arrigara, M. & Levina, N. 2008, "Social dynamics in online cultural fields", *ICIS 2008 Proceedings*. Available: www.researchgate.net/profile/Natalia_Levina/publication/221599714_Social_Dynamics_in_Online_Cultural_Fields/links/541af26a0cf2218008bff4e6/Social-Dynamics-in-Online-Cultural-Fields.pdf [2018, 2 July].
Ayton, P. & Tumber, H. 2001, "The rise and fall of perceived bias at the BBC: We should look to psychology not content analysis to understand allegations of political partiality", *Intermedia*, vol. 29, no. 4, pp. 12–15.
Bardoel, J. 1996, "Beyond Journalism", *European Journal of Communication*, vol. 11, no. 3, pp. 283–302.
Barnard, S.R. 2012, *Twitter and the journalistic field: How the growth of a new (s) medium is transforming journalism*, Doctoral dissertation, University of Missouri-Columbia.
Barnett, S. 2009, *Journalism, democracy and the public interest*, Reuters Institute for the Study of Journalism, Oxford.
Beck, U. 1992, *Risk society: Towards a new modernity*, Sage Publications, London.
Beckett, C. & Deuze, M. 2016, "On the role of emotion in the future of journalism", *Social Media+ Society*, vol. 2, no. 3, DOI:2056305116662395.
Bell, M. 1997, "TV news: How far should we go?" *British Journalism Review*, vol. 8, no. 1, pp. 7–16.
Bell, M. 1998, "The journalism of attachment", in *Media ethics*, ed. M. Kieran, Routledge, London, pp. 15–22.
Benkler, Y. 2006, *The wealth of networks*, Yale University Press, New Haven and London.
Benson, R.D. & Neveu, E. 2005, *Bourdieu and the journalistic field*, Polity Press, Cambridge.
Benthall, J. 1993, *Disasters, relief and the media*, I B Tauris, London.
Bergström, A. 2008, "The reluctant audience: Online participation in the Swedish journalistic context", *Westminster Papers in Communication and Culture*, vol. 5, no. 2, pp. 60–80.
Berry, M. 2013, "The today programme and the banking crisis", *Journalism*, vol. 14, no. 2, pp. 253–270.

Birrell, I. 2012, 18 February-last update, *Big charity: The aid business has grown fat* [Homepage of The Spectator], [Online]. Available: www.spectator.co.uk/essays/all/7652098/big-charity.thtml [2012, 22 February].

Boorstin, D.J. 1992, *The image: A guide to pseudo-events in America*, Vintage Books, New York.

Bourdieu, P. 1990, *The logic of practice*, Stanford University Press, Stanford.

Bourdieu, P. 1998, *On television and journalism*, Pluto, London.

Bourdieu, P. 2005,"The Political Field, the Social Science Field and the Journalistic Field", in *Bourdieu and the journalistic field,* eds. R.D. Benson & E. Ncveu, Polity Press, Cambridge, pp.29-48.

Brainard, C. 2010, 13 January-last update, *New media crucial in the aftermath of Haiti'* [Homepage of Columbia Journalism Review], [Online]. Available: www.cjr.org/the_observatory/new_media_crucial_in_aftermath.php [2012, 16 May].

Breed, W. 1955, "Social control in the newsroom: A functional analysis", *Social Forces*, vol. 33, no. 4, pp. 326–355.

Brighton, P. & Foy, D. 2007, *News values*, Sage Publications, London.

Broad, G.M. 2013, "Review: Robert W. McChesney, digital disconnect: How capitalism Is turning the internet against democracy", *International Journal of Communication*, vol. 7, pp. 1578–1581.

Brown, B. 2005, 16 January, "Then she pulled me closer and started to sob uncontrollably on my shoulder", *The Observer*, London.

Bunz, M. 2010, 14 January-last update, *In Haiti earthquake coverage, social media gives victim a voice* [Homepage of The Guardian], [Online]. Available: www.guardian.co.uk/media/pda/2010/jan/14/socialnetworking-haiti [2012, 16 May].

Burrell, I. 2005, 3 January-last update, *On the front line* [Homepage of The Independent], [Online]. Available: www.independent.co.uk/news/media/on-the-front-line-15131.html [2015, 8 January].

Carlson, M. & Lewis, S.C. 2015, *Boundaries of journalism: Professionalism, practices and participation*, Routledge, Oxford and New York.

CARMA. 2006, January-last update, *The Carma report on Western media coverage of humanitarian disasters* [Homepage of CARMA International], [Online]. Available: www.imaging-famine.org/images/pdfs/carma_%20report.pdf [2012, 24 April].

Chouliaraki, L. 2006, *The spectatorship of suffering*, Sage Publications, London.

Chouliaraki, L. 2010, "Ordinary witnessing in post-television news: Towards a new moral imagination", *Self-Mediation: Citizenship and New Media, Special Issue Critical Discourse Studies*, vol. 7, no. 3, pp. 305–319.

Cohen, S. 2001, *States of denial: Knowing about atrocities and suffering*, Polity Press, Cambridge.

Cohen, S. & Seu, B. 2002, "Knowing enough not to feel too much: Emotional thinking about human rights appeals", in *Truth claims: Representation and human rights*, eds. M. Bradley & P. Petro, Rutgers University Press, pp. 187–201.

Cooper, G. 2007, *Anyone here survived a wave, speak English and got a mobile? Aid agencies, the media and reporting disasters since the tsunami*, Nuffield College, Oxford.

Cooper, G. 2008, 20 May-last-update, *Disaster cap-it-all-ism* [Homepage of The Guardian], [Online]. Available: http://www.guardian.co.uk/commentisfree/2008/may/20/disastercapitallism [2018, 20 August]

Cooper, G. 2011a, *From their own correspondent? New media and the changes in disaster coverage: Lessons to be learned*, Reuters Institute for the Study of Journalism, Oxford.

Cooper, G. 2011b, "Why were women correspondents the face of coverage of the Libyan revolution?" in *Mirage in the desert? Reporting the Arab spring*, eds. R.L. Keeble & J. Mair, Abramis, London, pp. 236–245.

Cooper, G. 2015, 11 May-last update, *Nepal: Why are some disasters bigger than others?* [Homepage of BBC College of Journalism], [Online]. Available: www.bbc.co.uk/blogs/collegeofjournalism/entries/5e790c00-7ea9-4ee6-948d-3fee77cb26ce [2018, 28 April].

Cottle, S. 2000, "TV news, lay voices and the visualisation of environmental risks", in *Environmental risks and the media*, eds. B. Adam, S. Allan & C. Carter, Routledge, London, pp. 29–44.

Cottle, S. 2009, *Global crisis reporting: Journalism in the global age*, McGraw-Hill Education, New York.

Cottle, S. 2012, 29 February, *Demonstrations, riots and uprisings: Mediated dissent in a changing communication environment*, Public lecture given at the London School of Economics, London. Available: www.lse.ac.uk/website-archive/newsAndMedia/videoAndAudio/channels/publicLecturesAndEvents/player.aspx?id=1377 [2018, 28 June].

Cottle, S. & Nolan, D. 2009, 16 November-last update, *How the media's codes and rules influence the way NGOs work* [Homepage of Nieman Journalism Lab], [Online]. Available: www.niemanlab.org/2009/11/simon-cottle-and-david-nolan-how-the-medias-codes-and-rules-influence-the-ways-ngos-work/ [2015, 18 August].

Coward, R. 2013, *Speaking personally: The rise of subjective and confessional journalism*, Palgrave Macmillan, Basingstoke.

Cramer, C. 2005, 23 January-last update, *Once more with feeling*, Media Guardian edn [Homepage of The Guardian, London], Online. Available: www.theguardian.com/media/2005/jan/24/mondaymediasection11 [2018, 28 June].

Curran, J., Seaton, J. 2010, *Power without responsibility*, Routledge, London.

Daily Mail reporter, *Prince of Wales and Duchess of Cornwall pay tribute to victims of boxing day tsunami at opening of memorial*, 2012, 15 August-last update, [Homepage of Mail Online], [Online]. Available: www.dailymail.co.uk/news/article-2011983/Boxing-Day-tsunami-memorial-opens-Charles-Camilla-pay-tribute.html [2018, 20 August].

Davies, N. 2008, *Flat earth news: An award-winning reporter exposes falsehood, distortion and propaganda in the global media*, Chatto & Windus, London.

Davis, C. 2004, *Who IS right about MMR?* 2004, 11 March-last update, [Homepage of Mail Online], [Online]. Available: http://dailymail.co.uk/columnists/article-211412/Who-IS-right-MMR.html [2018, 21 August]

Davison, A. 2004, "Reinhabiting technology: Ends in means and the practice of place", *Technology in Society*, vol. 26, no. 1, pp. 85–97.

Deuze, M. 2007, *Media work*, Polity Press, Cambridge and Malden, MA.

Deuze, M. 2011, "Media life", *Media Culture & Society*, vol. 33, no. 1, pp. 137–148.

Dick, M. 2014, "Interactive infographics and news values", *Digital Journalism*, vol. 2, no. 4, pp. 490–506.

D'Zurilla, C. 2010, 12 January-last update, *Earthquake in Haiti: Celebrities and fans respond on Twitter* [Homepage of LA Times], [Online]. Available: http://

latimesblogs.latimes.com/gossip/2010/01/haiti-quake-celebrities-fans-twitter.html [2012, 7 March].

Earl, J. & Kimport, K. 2011, *Digitally enabled social change: Activism in the internet age*, MIT Press, Cambridge, MA.

Egeland, Jan speaking on ABC News, "The sympathy gap", *Nightline*, 2005, 11 January.

Eisensee, T. & Stromberg, D. 2007, "News droughts, news floods and US disaster relief", *Quarterly Journal of Economics*, vol. 122, no. 2, pp. 693–728.

Eley, G. (ed). 1992, *Nations publics and political cultures: Placing Habermas in the nineteenth century*, MIT Press, Cambridge, MA.

Evans, M. 2010, 15 January-last update, *Haitian earthquake dominates Twitter* [Homepage of Sysomos], [Online]. Available: http://blog.sysomos.com/2010/01/15/haitian-earthquake-dominates-twitter/ [2012, 17 May].

Fenton, N. 2010, *New media, old news: Journalism & democracy in the digital age*, Sage Publications, Thousand Oaks, CA, London.

Fortunati, L. 2002, "The mobile phone: Towards new categories and social relations", *Information, Communication & Society*, vol. 5, no. 4, pp. 513–528.

Franks, S. 2006, "The CARMA report: Western media coverage of humanitarian disasters", *The Political Quarterly*, vol. 77, no. 2, pp. 281–284.

Franks, S. 2015, "From pictures to policy: How does humanitarian reporting have an influence?" in *Humanitarianism, communications and change*, eds. S. Cottle & G. Cooper, Peter Lang, New York, pp. 153–167.

Fraser, N. 1990, "Rethinking the public sphere: A contribution to the critique of actually existing democracy", *Social Text*, vol. 25/6, pp. 56–80.

Freedman, D. 2012, "Web 2.0 and the death of the blockbuster economy", in *Misunderstanding the internet*, eds. J. Curran, N. Fenton & D. Freedman, Routledge, London & New York, pp. 69–94.

Freund, E. 2011, "Discuss this article!", *Participatory uses of comment sections on Spiegel online: A content analysis*, MSc Dissertation, LSE. Available: www.lse.ac.uk/media-and-communications/assets/documents/research/msc-dissertations/2010/Freund.pdf [2018, 2 July].

Frosh, P. & Pinchevski, A. 2009, "Crisis-readiness and media witnessing", *The Communication Review*, vol. 12, no. 3, pp. 295–304.

Fuchs, C. 2013. "Social media and capitalism", in *Producing the internet: Critical perspectives of social media*, ed. Tobias Olsson, Nordicom, Göteborg, pp. 25–44.

Galtung, J. & Ruge, M.H. 1965, "The structure of Foreign news: The presentation of the Congo, Cuba and Cyprus crises in four Norwegian newspapers", *Journal of Peace Research*, vol. 2, pp. 64–91.

Gans, H.J. 1980, *Deciding what's news: A study of CBS evening news, NBC nightly news, 'Newsweek' and 'Time'*, Constable, London.

Gieber, W. 1964, "News is what newspapermen make it", in *People society and mass communications*, eds. L.A. Dexter & D.M. White, Free Press, New York, pp. 173–182.

Gillmor, D. 2005, 6 January-last update, *Tsunami and citizen journalism's first draft* [Homepage of Dan Gillmor], [Online]. Available: http://dangillmor.typepad.com/dan_gillmor_on_grassroots/2005/01/tsunami_and_cit.html [2012, 16 May].

Gitlin, T. 2002, "10 Public sphere or public sphericules?" *Media, ritual and identity*, eds. J. Curran & T. Liebes, Routledge, London, pp. 168–174.

Golding, P. & Murdock, G. 1991, "Culture, communications, and political economy", in *Mass media and society*, eds. J. Curran & M. Gurevitch, 2nd edn, Edward Arnold, London, pp. 70–92.
Gowing, N. 1998, "New challenges and problems for information management in complex emergencies", *Disasters from Disaster Zones*, 27–28 May, p. 1. Available at: http://repository.forcedmigration.org/show_metadata.jsp?pid=fmo:1848 [2018, 28 June].
Gralnick, J. 2003, March-last update, *Lessons to survive by* [Homepage of TV Week], [Online]. Available: www.tvweek.com/news/2003/03/lessons_to_survive_by.php [2012, 17 May].
Greenwood, S., Perrin, A. & Duggan, M. 2016, 11 November-last update, *'Social media update 2016' Pew Research Center* [Homepage of Pew Research Center], [Online]. Available: www.pewinternet.org/2016/11/11/social-media-update-2016/ [2018, 28 April].
Habermas, J. 1989, *The structural transformation of the public sphere*, Polity Press, Cambridge.
Hall, S.C., Jefferson, C. and Clarke, T., J. and Roberts, B. 1978, *Policing the crisis: Mugging, the state, and law and order*, Palgrave Macmillan, London.
Harcup, T. & O'Neill, D. 2001, "What is news? Galtung and Ruge revisited", *Journalism Studies*, vol. 2, no. 2, pp. 261–280.
Harcup, T. & O'Neill, D. 2017, "What is news? News values revisited (again)", *Journalism Studies*, vol. 18, no. 12, pp. 1470–1488.
Hayes, A.S., Singer, J.B. & Ceppos, J. 2007, "Shifting roles, enduring values: The credible journalist in a digital age", *Journal of Mass Media Ethics*, vol. 22, no. 4, pp. 262–279.
Henderson, S., Taylor, R. & Thomson, R. 2002, "In touch: Young people, communication and technologies", *Information, Communication & Society*, vol. 5, no. 4, pp. 494–512.
Herman, E.S. & Chomsky, N. 1988, *Manufacturing consent: The political economy of the mass media*, Pantheon Books, New York.
Hermida, A. 2009, "The blogging BBC: Journalism blogs at 'the world's most trusted news organisation' ", *Journalism Practice*, vol. 3, no. 3, pp. 268–284.
Hesmondhalgh, D. 2006, "Bourdieu, the media and cultural production", *Media, Culture & Society*, vol. 28, no. 2, pp. 211–231.
Hoijer, B. 2004, "The discourse of global compassion: The audience and media reporting of human suffering", *Media, Culture and Society*, vol. 26, no. 4, pp. 513–531.
Jukes, S. 2017, *Affective journalism—uncovering the affective dimension of practice in the coverage of traumatic news*, PhD thesis Goldsmiths, University of London.
Katz, E. & Liebes, T. 2007, "No more peace! How disaster, terror and war have upstaged media events", *The International Journal of Communication*, vol. 1, pp. 157–166.
Kennan, G. 2003, 30 September-last update, *Somalia, through a glass darkly* [Homepage of New York Times], [Online]. Available: www.nytimes.com/1993/09/30/opinion/somalia-through-a-glass-darkly.html?pagewanted=all [2016, 21 March].
King, H. 2016, 7 October-last update, *People are livestreaming Hurricane Matthew* [Homepage of CNN.com], [Online]. Available: http://money.cnn.com/2016/10/07/technology/hurricane-matthew-live-stream-periscope/index.html [2018, 27 April].

Kitzinger, J. 1999, "Researching risk and the media", *Health, Risk & Society*, vol. 1, no. 1, pp. 55–69.

Klinenberg, E. 2005, "Channelling into the journalistic field: Youth activism and the media justice movement", in *Bourdieu and the journalistic field*, eds. R. Benson & E. Neveu, Polity Press, Cambridge, pp. 174–195.

Koch, T. 1991, *Journalism for the 21st century: Online information, databases and the News*, Greenwood Press, Westport, CT.

Krause, M. 2011, "Reporting and the transformations of the journalistic field: US news media, 1890–2000", *Media, Culture & Society*, vol. 33, no. 1, pp. 89–104.

Lasorsa, D.L., Lewis, S.C. & Holton, A.E. 2012, "Normalizing Twitter: Journalism practice in an emerging communication space", *Journalism studies*, vol. 13, no. 1, pp. 19–36.

Levinson, S. & Jamieson, A. 2017, 5 January-last update, *Four suspects charged with hate crimes over beating in Facebook live video* [Homepage of The Guardian], [Online]. Available: www.theguardian.com/us-news/2017/jan/05/facebook-live-beating-anti-donald-trump [2018, 27 April].

Liebes, T. 1998, "Television disaster marathons: A danger of democratic processes?" in *Media, ritual and identity*, eds. T. Liebes & J. Curran, Routledge, London, pp. 71–87.

Livingston, S. & Eachus, T. 1995, "Humanitarian crises and U.S. foreign policy: Somalia and the CNN effect reconsidered", *Political Communication*, vol. 12, pp. 413–429.

Livingston, S. & Eachus, T. 1999, "Rwanda: U.S. policy and television coverage", in *The path of a genocide: The Rwanda crisis from Uganda to Zaire*, eds. H. Adelman & A. Surkhe, Routledge, New York, pp. 209–228.

Livingstone, L. & Lunt, P. 1994, *Talk on television*, Routledge, London.

Loyn, D. 2005, 22 February-last update, *Covering the tsunami: A frontline discussion* [Homepage of DART Center], [Online]. Available: http://dartcenter.org/content/covering-tsunami-0 [2012, 16 May].

Macleod, L. 2010, 22 January-last update, *New media vital in breaking Haiti earthquake story* [Homepage of BBC], [Online]. Available: www.bbc.co.uk/worldservice/worldagenda/2010/01/100122_worldagenda_haiti_monitoring.shtml [2012, 16 May].

Manning, P. 2001, *News and news sources: A critical introduction*, Sage Publications, London.

Maton, K. 2008, "Habitus", in *Pierre Bourdieu: Key concepts*, eds. K. Maton & M. Grenfell Acumen, London, pp. 49–65.

McChesney, R.W. 2013, *Digital disconnect: How capitalism is turning the internet against democracy*, New Press, New York, NY.

McKenzie, H. 2008, 18 May, "My mercy mission into the cyclone zone with the Burmese heroes who put junta to shame", *Sunday Times*, London.

McLaughlin, L. 1993, "Feminism, the public sphere, media and democracy", *Media Culture Society*, vol. 15, no. 4, pp. 599–620.

Minear, L., Scott, C. & Weiss, T. 1996, *The news media, civil war, and humanitarian action*, Lynne Rienner, Boulder, CO.

Moeller, S.D. 1999, *Compassion fatigue: How the media sell disease, famine, war and death*, Routledge, London.

Moeller, S.D. 2006, "Regarding the pain of others: Media, bias and the coverage of international disasters", *Journal of International Affairs*, vol. 59, no. 2, pp. 173–196.

Morrison, D.E. & Tumber, H. 1988, *Journalists at war: The dynamics of news reporting during the Falklands conflict*, Sage Publications, Thousand Oaks, CA.

Murdock, G. 1973, "Political deviance: The press presentation of a militant mass demonstration", in *The manufacture of news production: A reader*, eds. S. Cohen & J. Young, Sage Publications, Beverly Hills, CA, pp. 206–225.

Myles, J.F. 2010, "Bourdieu-language-media", in *Bourdieu, language and the media*, ed. J.F. Myles, Palgrave Macmillan, London, pp. 9–22.

Nagar, N. 2011, *The loud public: The case of user comments in online news media*, Doctoral dissertation, State University of New York at Albany.

Neveu, E. 2005, "Bourdieu, the Frankfurt school and cultural studies: On some misunderstandings", in *Bourdieu and the journalistic field*, eds. R. Benson & E. Neveu, Polity Press, Cambridge, pp. 195–214.

Ofcom. 2017, June-last update, *Adults media use and attitudes* [Homepage of Ofcom], [Online]. Available: www.ofcom.org.uk/__data/assets/pdf_file/0020/102755/adults-media-use-attitudes-2017.pdf [2018, 29 April].

O'Neill, D. 2012, "No cause for celebration: the rise of celebrity news values in the British quality press", *Journalism Education*, vol. 1, no. 2, pp 26–44.

Pantti, M. 2011, "Disaster news and public emotions", in *The Routledge handbook of emotions and mass media: Routledge international handbooks*, eds. K. Döveling, C. von Scheve & E. Konijn, Routledge, London, New York, pp. 221–236.

Parker, N. 2008, "Burma's starving—but cops are busy . . . nicking Sun men", 15 May edn, *The Sun*, London.

Pennington, H. 2010, "The role of the media in public health crises", in *Risk communication and public health*, eds. P. Bennett, K. Calman & S. Curtis, Oxford University Press, Oxford, pp. 81–96.

Peters, J.D. 2005, *Courting the Abyss: Free speech and the liberal tradition*, University of Chicago Press, Chicago.

Phillips, A. 2015, *Journalism in context: Practice and theory for the digital age*, Routledge, London.

Philo, G. 1998, *The Zaire rebellion and the British media: An analysis of the reporting of the Zaire crisis in November 1996 and 1997 by the Glasgow media group*, Dispatches from Disaster Zones Conference, London.

Prentoulis, M., Tumber, H. & Webster, F. 2005, "Finding space: Women reporters at war", *Feminist Media Studies*, vol. 5, no. 3, pp. 374–377.

Rheingold, H. 2007, *Smart mobs: The next social revolution*, Basic books, New York.

Robinson, P. 2002, *The CNN effect: The myth of news, foreign policy and intervention*, Routledge, London.

Robinson, S. & De Shano, C. 2011, "'Anyone can know': Citizen journalism and the interpretive community of the mainstream press", *Journalism: Theory, Practice, Criticism*, vol. 12, no. 8, pp. 963–982.

Ross, S.A. 2004, *Towards new understandings: Journalists and humanitarian relief coverage*, Fritz Institute and Reuters Foundation, New York.

Russell, A. 2007, "Digital communication networks and the journalistic field: The 2005 French riots", *Critical Studies in Media Communication*, vol. 24, no. 4, pp. 285–302.

Schiller, D. 1979, "An historical approach to objectivity and professionalism in American news reporting", *Journal of Communication*, vol. 29, no. 4, pp. 46–57.

Schlesinger, P. 1978, *Putting reality together*, Constable, London.

Schlesinger, P. 1990, "Rethinking the sociology of journalism: Source strategies and the limits of media-centrism", in *Public communication: The new imperatives: Future directions for media research*, ed. M. Ferguson, Sage Publications, London, pp. 61–83.

Schlesinger, P. & Tumber, H. 1994, *Reporting crime: The media politics of criminal justice*, Oxford University Press, Oxford.

Schudson, M. 1989, "The sociology of news production", *Media, Culture & Society*, vol. 11, pp. 263–282.

Schudson, M. 2001, "The objectivity norm in American journalism*", *Journalism*, vol. 2, no. 2, pp. 149–170.

Schudson, M. & Anderson, C. 2009, "Objectivity, professionalism, and truth seeking in journalism", in *The handbook of journalism studies*, eds. K. Wahl-Jorgensen & T. Hanitzsch, Routledge, New York, pp. 88–101.

Schultz, I. 2007, "The journalistic gut feeling: Journalistic doxa, news habitus and orthodox news values", *Journalism Practice*, vol. 1, no. 2, pp. 190–207.

Seaton, J. 2005, *Carnage and the media: The making and breaking of news about violence*, 1st edn, Allen Lane, London.

Singer, J.B. 2005, "The political j-blogger 'normalizing' a new media form to fit old norms and practices", *Journalism*, vol. 6, no. 2, pp. 173–198.

Sterne, J. 2003, "Bourdieu, technique and technology", *Cultural Studies*, vol. 17, no. 3–4, pp. 367–389.

Taylor, P.M. 2003, "Journalism under fire: The reporting of war and international crises", in *News public relations and power*, ed. S. Cottle, Sage Publications, Thousand Oaks, CA, pp. 63–79.

Tierney, K., Bevc, C. & Kuligowski, E. 2006, "Metaphors matter: Disaster myths, media frames, and their consequences in Hurricane Katrina", *The Annals of the American Academy*, vol. 604, pp. 57–81.

Tryhorn, C. 2006, 25 August-last update, *Broadcasters veto Lebanon charity appeal* [Homepage of The Guardian], [Online]. Available: www.guardian.co.uk/media/2006/aug/25/tvnews.uknews [2012, 11 March].

Tuchman, G. 1972, *Making news: A study in the construction of reality*, Free Press, New York.

Tumber, H. 2014, "Back to the future? The sociology of news and journalism from black and white to the digital age", in *Media sociology, a reappraisal*, ed. S. Waisbord, Polity Press, Cambridge, pp. 63–78.

Tunstall, J. 1974, *Journalists at work: Specialist correspondents, their news organisations, news sources and competitor-colleagues*, Constable, London.

Van Belle, D. 2008, 29–31 May, "Agenda-setting and donor responsiveness to humanitarian crisis and donor aid (Paper 2.2 in The Role of News Media in the Governance Reform Agenda)", *Harvard-World Bank Workshop*, p. 1.

Vos, T.P. 2012, "New media, old criticism: Bloggers' press criticism and the journalistic field", *Journalism*, vol. 13, no. 7, pp. 850–868.

Wahl-Jorgensen, K. 2013, "The strategic ritual of emotionality: A case study of Pulitzer Prize-winning articles", *Journalism*, vol. 14, no. 1, pp. 129–145.

Wahl-Jorgensen, K. 2016, "Emotion and journalism", in *The Sage handbook of digital journalism*, eds. T. Witschge, C.W. Anderson, D. Domingo & A. Hermida, Sage Publications, London, pp. 128–143.

Wahl-Jorgensen, K. 2018, 2 March, "Challenging presentism in journalism studies: An emotional life history approach to understanding the lived experience of journalists", *Journalism*, published online. DOI:1011771464884918760670

Wahl-Jorgensen, K., Sambrook, R., Berry, M., Moore, K., Bennett, L., Cable, J., Garcia-Blanco, I., Kidd, J., Dencik, L. & Hintz, A. 2013, *BBC breadth of opinion review*, Cardiff School of Journalism, Media and Cultural Studies, Cardiff. Available: http://downloads.bbc.co.uk/bbctrust/assets/files/pdf/our_work/breadth_opinion/content_analysis.pdf [2018, 23 August].

Wahl-Jorgensen, K., Williams, A. & Wardle, C. 2010, "Audience views on user-generated content: Exploring the value of news from the bottom up", *Northern Lights: Film & Media Studies Yearbook*, vol. 8, no. 1, pp. 177–194.

Wakefield, A.J., Murch, S.H., Anthony, A., Linnell, J., Casson, D., Malik, M., Berelowitz, M., Dhillon, A.P., Thomson, M.A. & Harvey, P. 1998, "Retracted: Ileal-lymphoid-nodular hyperplasia, non-specific colitis, and pervasive developmental disorder in children", *The Lancet*, vol. 351, no. 9103, pp. 637–641.

Walter, J. 2006, December, *World disasters report 2006*, IFRC. Available: ifrc.org/en/publications-and-reports/world-disasters-report/wdr2006/ [2018, 2 July].

Wardle, C. 2006, "It could happen to you: The move towards 'personal' and 'societal' narratives in newspaper coverage of child murder, 1930–2000", *Journalism Studies*, vol. 7, no. 4, pp. 515–533.

White, C. 2002, "Open season on MMR", *British Medical Journal,* vol. 324, DOI: https://doi.org/10.1136/bmj.324.7329.120

White, D.M. 1950, "The 'Gate Keeper': A case study in the selection of news", *Journalism Quarterly*, vol. 27, no. 3, pp. 383–390.

Williams, A., Wardle, C. & Wahl-Jorgensen, K. 2011, "Have they got news for us? Audience revolution or business as usual at the BBC", *Journalism Practice*, vol. 5, no. 1, pp. 85–99.

Williams, R. 2013, *Raymond Williams on television (Routledge Revivals): Selected writings*, ed. A O'Connor, Routledge, Oxon.

Wright, K. 2014, *A quiet revolution: The moral economies shaping journalists' use of NGO-provided multimedia in mainstream news about Africa*, PhD submitted to Goldsmiths University of London.

Zelizer, B. 1993, "Journalists as interpretive communities", *Critical Studies in Mass Communication*, vol. 10, no. 3, pp. 219–237.

3 "The Odd Mucky Weekend, Not a One Night Stand"

The Relationship Between Aid Agencies and Journalists When Covering Disasters

> Y'know, [aid agencies and journalists] it's a mutually exploitative relationship. It's like two incredibly selfish people go to Brighton for the odd mucky weekend, mutually selfish, it suits them to do it and they both come back smiling – and when they both mutually feel the urge, it happens again. But it's not one-night stands; it's not as callous and using as that.
>
> (Interviewee 27, foreign editor, Sunday broadsheet)

When the devastating Kashmir earthquake struck on 10 October 2005, Dr Irfan Noor was at the Ayub medical and teaching institution in Abbottabad. Ceilings caved in, the emergency room had to be closed, but people continued to pour through the hospital doors. Dr Noor did all he could to help on the ground, but he also did something else – he emailed the BBC to tell them about the work he was doing and offered to write something about it for them.

The UGC Hub at the BBC liaised with Dr Noor, and he produced a 'Doctor's Diary' – a series of online accounts of his work in Mansehra, Odhi and Balakot – as he travelled to the North West Frontier as part of a medical relief team (Noor, 2005; Cooper, 2011). In itself, it was an interesting and, at times, emotional story of attempts to reach some of the most remote victims of the quake, but as an account of the aftermath of a disaster it was nothing out of the ordinary. Yet, it was significant: rather than needing a journalist on the ground or the intermediary of an aid agency to produce a case study, the BBC had found a first-person account.

Seven years later, Save the Children UK wanted to raise awareness of the West African hunger crisis. The stories of its victims had not had much media traction. So, instead of approaching the mainstream media for a traditional press trip, the #hiddencrisis campaign was planned as a Twitter event.

The organisation took Neal Mann—who was then Sky News digital media editor and had a huge Twitter following—out to Burkina Faso. Mann and Save the Children UK linked up with Storyful, the self-proclaimed 'world's first social media news agency', to plot their journey. There was not a sufficiently strong news peg to capture the attention of the mainstream media, but the trip generated interest in social media, because users could follow it in real time.

This interest was not always necessarily positive—particularly when James Ball, then a journalist at the *Guardian*, started a row on Twitter describing the trip as 'a disaster', 'Twitter tourism' and 'tasteless'.[1] But it showed how aid agencies could use social media tools to bypass the mainstream media, alter the conversation and interact with their audience.

The two examples above are significant examples of user-generated content and social media tilting the balance away from the typical NGO-journalist dichotomy. But how unusual are they? This chapter will look at the debates surrounding contemporary NGO–journalist relationships and the boundary (re)negotiations that have occurred between them—not just because of cost-cutting in news organisations and the professionalisation of NGOs, but because of external changes to their fields of activity. Through qualitative interviews with both mainstream media and aid agencies, it will examine how both articulate their current relationship, and the practicalities that influence it and look at what (if any) difference the advent of user-generated content has made to the source–media relationship in the reporting of humanitarian disasters.

Boundary (Re)Negotiations

Communication via the media is just one way in which aid agencies have traditionally approached a range of communication tasks. They produce information, reports and press releases to try to affect political agendas; they work closely with political players to try to influence both elite and public opinion; and they communicate directly with the public—often through fundraising—to increase awareness of their particular causes (Cottle and Nolan, 2007; Waisbord, 2011). Aid agencies usually have a pragmatic approach to engaging with the press, but there is frequently some internal conflict—for example, between fundraising, humanitarian teams and the media office—about the best way of communicating through the media and the kind of stories that need to be publicised.

NGOs are not all equally effective in their interactions with the news media; those working on 'non-political', non-controversial issues are more likely to receive media attention (Deacon, 2003). Different aid agencies have different newsmaking strategies and very variable resources, which will be described in greater detail later in this chapter.

Scholars have traditionally thought that rather than challenging journalistic norms, NGOs have sought to mimic journalists to maximise attention for their cause (Fenton, 2010; Cottle and Nolan, 2009), using the idea of 'media logic', as described by Altheide and Snow (1979). But Waisbord (2011) prefers to see NGOs' actions as part of the more widespread professionalisation of newsmaking in order to become 'news shapers' (Manheim,

1998). He puts forward the idea of 'journalistic' rather than 'media logic' as a better way of understanding the NGOs' approach—encompassing news values, media formats, working conditions and editorial positions. This may involve hiring former journalists, as pioneered by Christian Aid and Oxfam in the UK (Cooper, 2007, 2011) and Latin America (Waisbord, 2011), who then use tactics commonly employed by public relations or government agencies to shape the news agenda. For example, the UN High Commissioner for Refugees (UNHCR) put out a press release 'announcing' the millionth Syrian refugee in 2013—an attempt to get the ongoing refugee crisis on the news agenda by formulating a news peg (Edwards, 2013).

NGOs also try to maintain ongoing relationships with sympathetic reporters by working within this journalistic logic, as Waisbord puts it:

> It is impossible to characterize the relations between NGOs and journalists in terms of complete collaboration or opposition. Whereas some circumstantially develop good relationships, tensions are common.
> (2011:151)

This tension—and collaboration—has increased with the cuts to foreign reporting. As far back as 2009, the think-tank POLIS and Oxfam published a report warning of the risk that international coverage by public sector broadcasters was likely to decrease (Harding, 2009). The previous year, the *Daily Mirror*—the newspaper that broke the story of the killing fields of the Khmer Rouge in 1979—announced that, as part of a round of job cuts, it was dispensing with the role of foreign editor entirely (Brook, 2008). In fact, the Media Standards Trust report found a 40 percent drop in the number of overseas stories in newspapers between 1979 and 2009 (Moore and Loyn, 2010). These developments form what Ulf Hannerz calls the 'paradox' of current international news: "In an era of intense globalisations, foreign news coverage in many media channels has recently been shrinking" (2012:23).

Into this breach have stepped two new potential players in the field. First there are 'accidental journalists' as detailed in Chapter 7. But this chapter looks at a new player—the aid agency as 'journalistic entity' (Powers, 2015). With the increasing professionalism of aid agencies' press operations and their increasing use of film, photographs and reports accessible to the mainstream media, aid agencies are undoubtedly now players in the journalistic arena.

Powers (2015) makes a valuable point when he says that the content produced by non-governmental organisations is expanding the boundaries of journalism—by which he means providing information about various parts of the world. This is happening in spite of the fact NGOs themselves are not driven by journalistic instinct *per se*, but to seek legitimacy by providing information to political elites via reports and newsworthy coverage. I would contend, however, that international aid agencies are not solely driven by the need to influence political elites, but also by their desire to attract business donors (see Save the Children's co-operation in 2011 with GlaxoSmithKline for example).[2]

Journalists are increasingly incorporating NGO content into their coverage—for example, linking to reports or videos (Wright, 2014, 2016, 2018)—while still seeing their work as separate from what NGOs do. Over certain stories, such as climate change or humanitarian disasters, however, there is often a sense of coming together to 'co-produce' the news. This is particularly evident when TV reports of an emergency are followed up with the details of an appeal or in national newspapers' annual Christmas appeals.[3]

In some cases, aid agency footage or text is used as a direct substitute for journalistic reporting (Cooper, 2007, 2011; Wright, 2014, 2016) although many journalists still argue against using this content, and also in interviews for this book disputed how widely it is used.

But, like others (Cottle and Nolan, 2007, 2009; Fenton, 2010), Powers argues that NGOs are still subject to journalistic norms and practices rather than subverting them, and while interviews with aid agencies suggested some weariness about media norms and practices, there was little attempt to change such norms.

The fact remains, whether journalism was the initial impetus or not, the field has still been altered by the entry of NGOs. The question is, then, whether UGC and its creators can be a significant third player in the journalistic field, or whether it remains a duumvirate.

Aid Agencies and Journalists Working Together in the Field

Unlike the crime beat or health reporting—few journalists are specialists in reporting disasters. Those interviewed tended to be what are known in old-fashioned newspaper gendered parlance as 'firemen' or 'parachute reporters'—reporters dispatched at a moment's notice when a big event happens and who have experience of arriving in a foreign country and quickly assessing a story. An established model is employed—numbers dead, comparison to other disasters, emotional accounts with survivors—in what is known as template journalism.[4]

Such reporters are often experienced, but in the absence of regular bureau staff, the nuances of individual countries or regimes may be lost. With little time to prepare, and with frequent difficulty gaining access, such journalists may also tend to rely on trusted sources such as aid agencies.

Thus, one of the first questions for the editors interviewed was whom they commonly sent to cover humanitarian stories—foreign correspondents (based in country or nearby), firemen, general reporters or freelances.

The financial strain most media outlets are under meant very few mentioned foreign correspondents, except in the case of the Haiti earthquake, which was near to US bureaux. Firemen/parachute reporters, or a senior correspondent, were most commonly mentioned.

A foreign editor at a daily broadsheet, recalling his experience during the Haiti quake, said it was important not to send "someone green" (Interviewee 24)—the exact phrase used by the head of international news at a commercial broadcast organisation, who typically deployed a number of foreign correspondents now based in the UK for such stories, because:

> If you send somebody who's really green; they'll turn up in a refugee camp and be utterly shocked, not realising that that is actually kind of status quo around the world and has been for a very long time, which is in itself is shocking but is not new.
>
> (Interviewee 39: head of international news, commercial broadcaster)

The editor of a flagship news programme on a public service channel (Interviewee 35) spoke of the need for seasoned presenters on a huge story. He sent out his two most experienced presenters to cover the Great East Japan quake because:

> The two of them compete voraciously, and they're really, really experienced. It was a very gruelling story because of the time it takes to get to Japan, and also the time difference, and they were both of them reporters who won't stop until they get the story.

The foreign editor of a daily broadsheet also stressed the need for experienced journalists if the story was to be told as quickly and accurately as possible. He sent one of his most senior correspondents to Haiti, for example.

> Haiti was the biggest one, and it was one of these tricky ones. With a humanitarian crisis there's a logistics thing of getting your people there, which is always costly and potentially quite dangerous, and in particular somewhere like Haiti where there's not really much infrastructure and order, so you've got to plan quite carefully whom you're going to send out. You want to make sure they're people of experience, and there's an awful lot of politics kicks in immediately. So you need someone who can navigate themselves quite artfully through that while bringing back the story but also making sure that they don't fall into some of the bad traps that come along.
>
> (Interviewee 19)

This need for 'experience' and careful planning, however, that the editors all stressed was not reflected by the reporters—many of whom were senior now, but had not been when they were first sent out on disaster stories. The emphasis from the ground was much more chaotic. One reporter said he was sent to cover humanitarian crises early in his career simply because he "was willing to go" and that, as a general reporter, he was qualified to write "sob stories about people burying bodies" (Interviewee 25, foreign correspondent for a broadsheet). Another writer for the same paper was sent out to a disaster simply because s/he had been on holiday nearby at the time. Another broadsheet writer was sent not because s/he was nearest—the paper had an experienced stringer nearer—but s/he was a staff writer and therefore it would cost the paper less. Meanwhile, another who had worked for both major daily midmarket papers said she owed her 'fireman' career to the fact that she knew several languages, however (in)appropriate they were (Interviewee 31).

She also pointed out how difficult it was to prepare for a humanitarian story, because there was rarely more than 24-hours' notice: "There's never any planning. It's all on the hoof. You might be able to buy yourself a little time because you've got to get a visa, and embassies, of course, work at their own pace although they can be induced to work quicker". The international editor for a commercial broadcaster said that he had files full of "yellowing news cuttings" that he kept on major subjects so that his background knowledge was robust and often because he only started researching the subject once he was getting on the plane (Interviewee 38). One broadsheet correspondent covering the East Africa drought only had time to quickly email a few cuttings to him/herself to read them on his/her laptop on the plane (Interviewee 28).

This was not true for everyone. A foreign affairs editor at a satellite broadcaster said most of his trips were pre-arranged for slow-onset emergencies while a foreign correspondent for a tabloid said that around half of his trips were planned (Interviewee 40, Interviewee 33). That was because the tabloid reporter was sent out to write longer features about the aftermath of a disaster rather than breaking news.

It is not just the background reading that may be sketchy. Because of the unpredictable nature of humanitarian disasters, journalists are often dispatched at short notice, so transport, fixers and translators tend to be organised on a very *ad hoc* basis. When covering his first disaster—the 1999 Izmit earthquake in Turkey—a foreign correspondent for a daily broadsheet recalled recruiting a waiter at his hotel as his translator, on the basis that the waiter spoke good English, had a car to transport them around and, because he was working nightshifts, was available during the day (Interviewee 25).

Journalists may also have difficulty getting to the scene of a disaster without help from an external agency such as the UN, government agencies or international NGOs. For example, a senior emergencies specialist at a large INGO recalled being stuck in the Dominican Republic after the Haitian earthquake, trying to negotiate with the Department for International Development (DFID) to get journalists onto a plane into Port-au-Prince (Interviewee 18). Similarly, a senior BBC correspondent remembered sitting in a tiny restaurant in Darfur surrounded by staff from five of the big aid agencies—including Oxfam, the World Food Programme and Save the Children—all vying to offer her a seat on a helicopter (Cooper, 2007).

Why does this matter? The difficulties faced by unprepared journalists working at short notice mean they may find themselves depending on alternative sources of information and guidance—and more easily influenced. For some journalists, this has gone further than informal dependence: whole trips have been 'sponsored' by NGOs, a phenomenon described as 'beneficent embedding' (Cottle and Nolan, 2009).[5] This trend shows "spheres are overlapping more and more" (Roberts cited in Abbott, 2009).

To tease out the state of the source–media relationship in these stories, I asked a series of questions to both journalists and aid agencies to try to discover how dependent journalists in a budget-cutting environment were on aid agencies, in terms of getting to the story and reporting on it. I then compared the two different accounts.

The Journalists' Account

When does taking help from an aid agency compromise journalistic independence? Where is the line between taking the help a journalist needs to get the story and becoming dependent on an organisation with its own agenda? Out of the 18 editors and reporters who answered the questions about what they accepted from aid agencies, there were only three (two newspaper journalists and one broadcast journalist) who said they would not accept any help at all. Even these few instances of claiming to refuse help were contradicted, however—in two out of the three cases another reporter/editor from the same organisation said that the organisation had accepted help from an agency at times.

Journalists were most likely to accept transport *in situ* and accommodation, seeing this as a practical response to working in difficult circumstances. They were warier of using fixers/translators or flights provided by aid agencies.

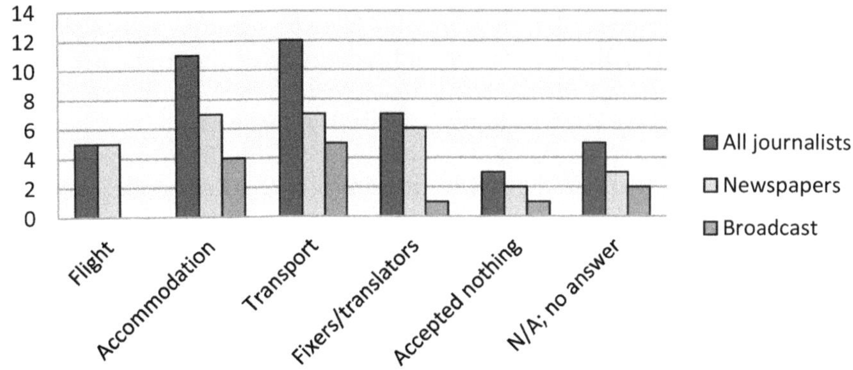

Figure 3.1 What Journalists Say They Accept from Aid Agencies

There was a 'sliding scale': most journalists were reluctant to say that they would use aid agencies' facilities for fear of appearing compromised but accepted that sometimes they would need their help to get around. For example, Interviewee 38, the international editor for a commercial broadcaster, said that he would sort out his own fixer and translator, but that if there was no other way to get to a particular story then he would go on an aid agency trip. He believed, however, that the rise of UGC meant it was becoming far less likely that journalists would need to do so.

> If there's something happening in Kenya or Ethiopia, we should know about it as well. It may well be the aid agencies on the ground who are first ringing the alarm bells, and that's good, but we also have staff in Nairobi or wherever, whom we also pay to ring alarm bells for us. Therefore, I think the days when you were in the dark and an aid agency said, "You've got no idea what's about to happen in Sudan, we've got transport, a plane, do you fancy joining?" have gone. The world has changed, for a lot of the reasons that we've been talking about: user-generated content. I'm not saying that [someone] in Sudan is managing to get on YouTube about conditions in his local village, but it's not impossible now, whereas 20 years ago it simply would have been.
>
> (Interviewee 38)

His editor did not completely rule out sharing fixers or translators or taking flights from an aid agency, but said that he would be extremely reluctant to take anything from any outside agency, whether charitable or commercial, because of maintaining "independence" and "integrity" (Interviewee 37). He did go on to add, however, that if an NGO was helping them to get to a location and offered them a lift then it would be "insane" not to take advantage.

The fact that journalists have become a target for armed groups and terrorists in recent times (Cooper and Cottle, 2015) meant that some editors saw closer co-operation with aid agencies being less to do with journalistic independence and more to do with security. Interviewee 34, who worked for a tabloid, mentioned that her paper had become much more security-conscious after the deaths of reporter Rupert Hamer and wounding of photographer Philip Coburn in Afghanistan.[6] Meanwhile a broadsheet writer had heard reports of hijackings along the road to Kenya's Dadaab refugee camp. She accepted protection from Unicef, a large and established NGO in the area, which was able to provide her with safe transportation and secure accommodation in the camp itself—something that the paper could not afford (Interviewee 28).

There was also the matter of using NGOs to get to places that would be not have been accessible otherwise. The foreign editor of a daily broadsheet said that he would send one of his reporters on a trip [with an aid agency] possibly once a year, but this was about being able to tell a story that would not otherwise have been told:

> If they're laying on a trip to a very remote area that realistically we're just not going to be able to get to, or afford to get to, and then they'll give us an insight into what's going on, then I'm very happy to do that. But I don't think we ever commit ourselves to say, "We will give you three stories" or something.
>
> (Interviewee 19)

There was some conflict between editors and reporters over what would and would not be accepted. For example, the head of international news at a commercial broadcast organisation (Interviewee 39) said she would be happy for NGOs to sort out transport and accommodation if it was a question of security or they were better placed to do so. But a foreign affairs editor at a satellite broadcaster was adamant that he would not accept anything from NGOs although this reflected his senior position, which meant he was much more likely to take planned trips than be parachuted in (Interviewee 40).

Similarly, an editor said his programme's policy was to never allow a charity to pay for anything.

> Even if we were embedded with the UN, I'd always say "how much do you think the board and person costs?" . . . I mean, ethically you don't want an aid agency to be paying for you when they should be giving the money to kids—and you also don't want a story coming out [about] that, y'know.
>
> (Interviewee 35, editor, flagship news programme, public service broadcaster)

Yet his international editor said that in certain circumstances she would accept help from an aid agency, even though she'd much rather remain independent and pay her own way. She cited a story she'd covered on mass rape in St Lucia, when she had accompanied the UN special representative on sexual crime in conflict to meet the women in question and accepted UN support.

> Now that's an occasion where I did take the transport and the accommodation. . . . But, you know, without them we couldn't have done that story, and I think that was a really important story to do. That was really the first time that people had seen what those women had been through, had heard those stories, had heard her [the UN representative's] reaction, and then because of that, there were arrests made. That, to me, was an example of where it's absolutely the right thing to do. But I don't think it's the right thing to do if you have an alternative and if it's not a news story, basically if you're being lazy or if you're sponging off the aid agencies.
>
> (Interviewee 36)

There is a complex negotiating system that can come into play as well: one photographer offered to give an NGO photographs he was taking, in return for a lift into a refugee camp. And when one broadsheet reporter needed translators, she asked Oxfam for suggestions, and they gave her lists of local people. "They were just people living in the town who lots of people different organisations and journalists had used, so I don't think there was any kind of agenda there" (Interviewee 28).

Tabloid newspapers—who were particularly cash-strapped—were more willing to admit using aid agency facilities. A foreign correspondent at a tabloid said that Plan International had paid for his flight to cover the Niger food crisis of 2005, and had done so on other occasions too.

> I'd say we have pretty good relationships with them [aid agencies]. Of course they love to be involved with us in any way that they can, and they always help us as much as they can. . . . For example, I just had a conversation over the weekend with our reporter in Damascus, and she happened to mention that she was seeing Unicef because we were having trouble getting access to the heartrending human interest stories that we ought to be seeing in Syria, because we've gone in there officially with a visa, so our reporter is working under certain constraints. She can move around but she's not necessarily going to be given free access to injured people in hospital, she can't go across the frontlines into the rebel areas where the victims of artillery and air strikes are going to be. So we're trying another way, through the aid agencies. And if Unicef can lead us to the stories of children that they've worked with and that we can photograph and that we think will produce a harrowing narrative then we'll take that.
>
> (Interviewee 33)

A feature writer and 'fireman' reporter for a midmarket paper said she saw foreign trips disappear in recent years because of changes both in budgets and in newsroom priorities (Interviewee 31). She said, however, that—even in the past—there had been a reliance on aid agencies to foot the bill, although she personally disapproved of it. She talked of two trips—one with Save the Children UK to Cambodia and one with Oxfam GB to Uganda—on which the NGOs had paid for everything. Meanwhile, for the most hard-up quality newspapers, the budget for foreign coverage was under such stress that the only way to cover stories was by taking NGO facilities, as one foreign editor of a broadsheet explained.

> I'm afraid our deal was pretty much that we would always accept stuff like that. That's not the purest view but there were many, many times when we couldn't go somewhere. I probably would say they would offer those things on an understanding that there would be a piece and I would always try to do something. I don't feel as if I would ever

> have let that dictate the kind of content that we did. But I suppose you always think that, don't you—and perhaps it did. You know if those offers stopped coming, the truth is it would affect the quality of our foreign coverage. . . . So I think it is a bit of a dodgy deal and I also think it's important you apply the same standards to them [the aid agencies] even though they're the good guys. You can't just switch off your fantasies about whether it's going to affect what you're doing but, ultimately, I just think our readers would be getting a worse deal if we didn't do it.
> (Interviewee 24: foreign editor, daily broadsheet)

The foreign editor on the Sunday sister paper said that he had seen a 'revolution' in the relationship between aid agencies and newspapers during his time as a journalist (Interviewee 27). For him, the increasing squeeze on foreign budgets meant that newspapers had increasingly taken aid agencies' offers.

Many reporters were definitely concerned that their journalism could be affected by 'beneficent embedding' (Cottle and Nolan, 2009). Interviewee 28 was sent out by her broadsheet newspaper to cover the East Africa famine only to find the story that the aid agencies wanted her to cover was not as good as the one she discovered. In the end she had to compromise.

> I flew out, we paid for the flights, though I think they [the editors] probably would have preferred not to, but nothing could be arranged fast enough. So I got to Nairobi and I was trying to get to the refugee camp, which was quite hard to do and incredibly expensive to do safely. So the first thing I had to do was cadge a lift with somebody. Again, the obvious choice is somebody like Unicef who have the infrastructure to drive you there. I'm fascinated by it because when I look back at this trip I find it actually quite perverse how much you end up embedded, effectively. It's no different to embedding with an army, is it? . . . I'm sure there are other people in this position, but as a young journalist very keen do to foreign reporting, that's why you persuade your editor to let you do it. It'll be zero cost involved for them. But it's hard. Obviously you're kind of aware that there are parts that could be hugely compromising.

The Aid Agencies' View

How did the perspective of aid agencies compare with what the journalists revealed? Did they have the same view of the relationship?

Comparisons are difficult because of the numbers involved (there were 13 aid agency representatives able to answer these questions compared with 18 journalists). But Figure 3.3 gives a rough sense that the NGOs generally report providing more for journalists than journalists admit accepting. This may be explained by the fact that NGOs are very focused on getting

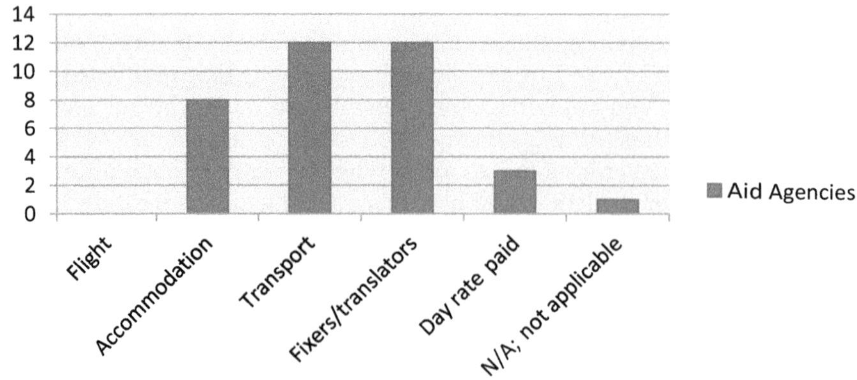

Figure 3.2 What Aid Agencies Say They Provide for Journalists

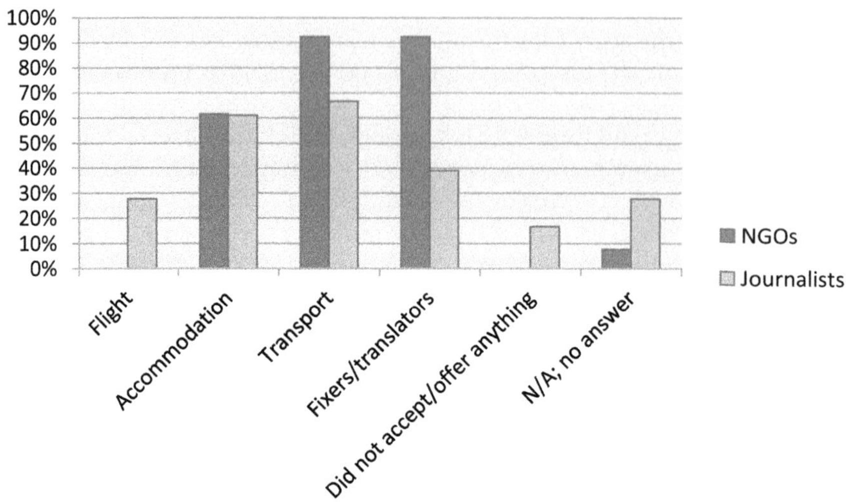

Figure 3.3 Comparison of What Journalists and Aid Agencies Say is Accepted

media attention, whereas for journalists, NGOs were an important source in humanitarian crisis stories, but only one of many. It may be that faulty memories were clouding what actually happened. Or it may be that journalists were reluctant to admit how dependent they were on NGOs in the field.

The discrepancy in the answer over accepting flights (the one where journalists admitted accepting more than NGOs said they provided) can be explained as follows. Some of the NGOs did admit providing flights but this tended to be in very specific circumstances—and, interestingly, no NGOs talked about giving flights to national media journalists. It was the source of some tension, because while the larger agencies had the clout *not* to pay for anything, some of the smaller agencies in more desperate need of exposure—either for themselves or a disaster they wanted to highlight—were willing to pay for flights and freelancers.

> We are happy to host journalists, which means accommodation, drivers, interpreters, and we do sometimes pay for a flight for them, but we're finding that the BBC won't let them do that anymore. The last time we took Sky somewhere, they paid part of it because they need more people for telly, so it isn't like taking one radio person and comping a journey, and all that. Obviously we don't pay them,[7] but we would pay all of the costs relating to the trip, or as much of it as they wanted us to.
>
> (Interviewee 17, Head of Media and Corporate Communications, mid-sized aid agency)

For other agencies there was a division between what they were willing to provide for national and regional media. For example, one aid agency was willing to pay for flights—and accommodation, fixers and transport in country—for cash-strapped regional papers but not for national papers. NGOs who employed former journalists in their press offices tended to be more reluctant to pay for journalists. At two faith-based agencies where the press offices were dominated by ex-journalists rather than traditional PRs—there was a sense that if a media organisation had to pay part of the costs, then they were more likely to commit to a story, and also satisfy their ethical concerns.

> Personally, I think we shouldn't pay, but not because of the financial outlay but because if the news organisation hasn't paid out, then it hasn't reached the bar of being commissioned. They've got to put their money where their mouth is. So you [the agency] can pay for a trip, and then the story sits on the spike or on the 'almost-spike' for ever because there isn't a peg for it. But if a commissioning editor has taken it out of their budget, then they'll get it into the paper.
>
> (Interviewee 18, senior emergencies specialist, large aid agency)

> There's also the whole question from the journalist's point of view of impartiality. And, while it would be quite easy to defend using [our agency] vehicles and [our agency] staff, it probably wouldn't be that easy to defend having had your trip paid for you.
>
> (Interviewee 8, head of media, large aid agency)

For items such as accommodation, transport or food, it was simply a question of practicalities, with the added benefit that the journalists would focus on their work. There was a clear—if unspoken—*quid pro quo*: access to case studies, a camp or a remote area meant that there would be an interview with or a name-check for the aid agency. For many aid agencies, it was simply a logical progression to offer their own staff as translators or fixers, which had the added bonus for the journalists that there were no further costs on top.

> Typically, our fixers are local staff. We employ locals in country. They pretty much know the situation so they can act as fixers.
>
> (Interviewee 14 news editor, mid-sized aid agency)

> [We] wouldn't hire separate translators or fixers but, in practice, the local staff are usually bilingual and know enough about the local situation to 'fix' in an informal way. [We] have 46,000 staff after all. There are local 'communicators' in most countries who act as translators/fixers when needed for visiting international journalists.
>
> (Interviewee 18, senior emergencies specialist, large aid agency)

A 'fixer', in journalistic parlance, refers to someone who works for a journalist and helps by arranging interviews; conducting basic reporting; sorting out background briefings, security assessments and guidance; establishing local contacts with other media staff to help put together the reporter's story; and either facilitating a translator or translating themselves (Hamilton and Jenner, 2004; Palmer and Fontan, 2007). Often overlooked in academic studies, fixers can be a valuable source in directing journalists in where to go and whom to speak to. As such, they are not just there to sort the logistics but play an editorial role as well (Murrell, 2010).

In terms of field theory, as Murrell (2010, 2013) points out, fixers do not possess all three types of cultural capital Bourdieu identifies in foreign correspondents: embodied (have undergone self-improvement through education or training), objectified (deliver output—e.g. broadcast/printed news stories) and institutionalised (receive recognition through prizes and diplomas) (Murrell, 2010:129). Embodied cultural capital would seem to be the only one of the three they can claim. Yet, the increasing use of fixers by foreign correspondents means that they have become a powerful and vital part of the relaying of information between event and journalist, with the risk that without in-depth knowledge and context by the reporter, the event is seen through the filter of the fixer (Palmer and Fontan, 2007).

Some interesting dynamics are created if these fixers are employed by the aid agency. On one level it may merely mean the fixer is someone the journalist feels comfortable with—'people like us' as Murrell (2013, 2014) puts it. But this is a very particular filter to put on. Even if they do not control the final copy—and the aid agencies say they do not ask to see copy in general—by facilitating this sort of help, it could mean the story is framed in the way that most appeals to the agency.

> I wouldn't necessarily expect them [the journalist] to say this was a [name of INGO] project, but I would expect them to interview someone from here. . . . A key part of my job is ensuring that [our organisation] gets a credit, so that's my priority.
>
> (Interviewee 18, senior emergencies specialist, large aid agency)

At best, for the aid agency this can mean the reporter sees everything through the prism of its work.

> When I started at [my agency], it was the 2005 Malawi floods and I went with a 5 Live reporter . . . and literally she stayed with us throughout the whole of the trip, from beginning to end. I picked her up from the BBC, she got into the taxi, we went to the airport, we arranged the accommodation,[8] the drivers, we had our partners that told us which communities we needed to go to which were worst affected. She had her kit, and she filed her stories from the ground into various 5 Live programmes, and we were very much living in each other's pockets. She relied on us to get to remote places, and we relied on her to be able to impart what was happening on the ground very immediately.
>
> (Interviewee 6, head of World News Unit, aid agency)

The lines got even more blurred in the use of freelancers. While editors often had fairly clear policies when it came to what they would and would not allow their staff to accept, when it came to freelance submissions there tended to be less oversight. One interviewee, who worked for tabloids, said she had looked into being funded by NGOs for her freelance work, despite the ethical difficulties involved, on the grounds that stories on humanitarian crises risked becoming 'vanity' projects because of their expense.

> Last time I went to South Sudan it ended up costing me money—[I] had a magazine cover but was paid less than [the] cost of flights, warzone insurance and ultimately lost two weeks' normal work, plus the huge childcare cost. That's why I don't do it very often now, I can't afford vanity journalism and I have mouths to feed.
>
> (Interviewee 34)

Many agencies preferred to use staff reporters because there was more chance of the stories seeing the light of day, but others would use freelancers in a number of different ways in order to get a story into a newspaper, on the radio or, rarely, on television.

For example, some agencies would not usually pay for flights, but would make an exception for what they called "starving freelancers"—who would often receive little or no help from the newspaper with the expenses at all. A senior emergencies specialist recalled taking a freelancer to Nicaragua, who had had to subsidise the trip as well (Interviewee 18). Some agencies would go further and pay the freelancer a day rate—or pay if the freelancer was covering a celebrity visit to one of its projects.

> It's always been fairly successful because freelancers can get different stories in different places, but when we do that we always ask them to confirm that they have at least one outlet booked in.
>
> (Interviewee 2, head of news, aid agency)

Another large aid agency also used freelancers in a way that often required negotiation with newspapers. Its senior humanitarian news officer recalled taking freelancers for the *FT Magazine* to Cambodia and a freelance radio journalist to Gaza, as well as an illustrator, Olivier Kudlow, to Burkina Faso (Interviewee 13).

With the Burkina Faso trip, he said he wanted some assurance from the *Guardian* that it would be happy to take the drawings Kudlow produced during the trip and that would pay a token fee for them:

> We'd pay him [Olivier] for whatever amount of time he was there—five, six, seven days—and then we'd say: "We've agreed with the *Guardian* that they'll pay you a fee, if that fee falls short of what you would normally expect, we will subsidise it." But it's the *Guardian*'s responsibility to pay freelancers. They're a paper, and we're not undercutting freelancers.
>
> (Interviewee 13)

While many aid agencies would not pay for freelance journalists, several were happy to employ freelance photographers to go out into the field, then offer the photographs to newspapers for free, or allow the photographer to sell them to the outlet. Others would pay an approved freelance to go on a trip and offer their pictures to a media outlet for free, but the photographers were not allowed to sell them on. One said her aid agency preferred to be in charge of the pictures:

> Where we're doing a journalist trip—especially a print one—we obviously have to talk to the picture editor and, if they don't like the photographer we're sending, we have to find a different one. But we do prefer to send a photographer that we are paying rather than them send a photographer, because you know what it's like when you do a field trip and people are all over the place and trying to do something that you think, "oh, no, please don't go in that person's house", or whatever.

> It just means that you've just got more control—if you're paying the photographer, you're in charge, and then the photos are yours as well.
> (Interviewee 17, head of media and corporate communications)

Writing the Story: Aid Agency Content and Journalists' Use of It

Going beyond the idea of 'beneficent embedding' is the use of aid agency content and how this is labelled for the reader/viewer. From the late 1990s onwards, many aid agencies began employing journalists in the press office. Christian Aid hired journalists from the Press Association and the *Sunday Times* to try to improve their media profile. Oxfam followed suit, hiring a former Reuters bureau chief (Cooper, 2007). Both agencies reformulated their press offices to more closely resemble media 'newsrooms' and used their press officers in a similar way to 'fireman reporters'—putting them on the ground as soon as possible after a disaster and expecting them to produce their own copy and film as well as facilitate media requests.

By 2013, the bigger, more high-profile agencies were the ones employing most journalists and by 2018 at least half of the press office staff at Oxfam, World Vision, Christian Aid, Action Aid and the British Red Cross in the UK head offices were former journalists. Figure 3.4 on the next page reflects the media contacts listed on all their websites and how staff identified their team or self-identified themselves either on the websites or via social media profiles such as LinkedIn or Twitter. They had not all necessarily former foreign correspondents or Channel 4 News editors though over the years. They had included over the years a former producer at *Daybreak*, ITV's morning show, and a writer from *Take A Break* magazine: "There was a feeling that folksy stuff could be really useful for us, rather than doing drab development articles, and trying to make it a bit more lively" (Interviewee 7, world news officer).

Interviewee 10 said that—particularly at his level—journalistic experience was very useful. "I think it's not essential but it's the kind of poacher and gamekeeper relationship, and if you don't have some kind of handle on who's on the other side of the fence, you won't be good at the job".

This was even reflected in some press officer job titles—'News Editor', 'Head of News', 'Head of World News Unit'—to encourage a sense of journalistic professionalism. Ironically, one of the agencies which had led the field by calling its press officers 'journalists' reverted to more traditional job titles in 2011/12 as part of a restructure and as a reaction to the perceived power that the press office had wielded.

> There is an antipathy in this organisation to journalists, and that antipathy is the result of one stage—when the press office was a kind of semi-autonomous republic, if not indeed a completely autonomous republic.

> Unfortunately, all the people who were told to fuck off or whatever [by the press office] at that time are now in positions of considerable authority here. So I've inherited the payback.
>
> (Interviewee 8, head of media)

The use of former journalists at aid agencies was welcomed by many journalists because they felt their requests would be more rapidly understood and acted upon—sometimes at the expense of other, less media-savvy organisations:

> You can usually tell with the NGO's communications team who's been a journalist, because they're very good at coming up with the 'line', facts and a case study. You know that Save the Children is always going to have a good story for you in any disaster, because a third of all victims are kids, and there's always going to be a headline you can write saying millions, thousands of kids at risk, blah blah blah, and they provide it. I think other, perhaps lesser-known organisations doing good stuff, probably don't get the coverage they deserve.
>
> (Interviewee 25, foreign correspondent, daily broadsheet)

But the new breed of press officers went further than just fielding journalists' requests; they were now capable of writing or filming their own stories and offering them to media outlets who lacked the kind of staffing levels

DEC agency	*No. of staff in media team June 2018*	*No. of former journalists in media team June 2018*
Action Aid	4	1
Age International	1	0
British Red Cross*	8	4
CAFOD	6	4
CARE International	3	1
Christian Aid*	7	5
Concern	2	1
Islamic Relief	4.5***	1.5***
Oxfam GB*	11	7
Plan UK	4	2
Save the Children UK*	8**	5
Tearfund	9	3
World Vision UK	4	2

Figure 3.4 Employment of Former Journalists in DEC Member Press Offices, 2018

* These agencies have regional press officers as well, but only the national ones are counted for the purpose of this table

** This only counts Save the Children's news team, not their PR and celebrity teams

*** Full-time equivalents

and resources they had in the past. A key example of this occurred during Cyclone Nargis in Burma.

On 18 May 2008, the BBC's *News at Ten* led with a remarkable account of the cyclone. The film package told the poignant story of one family left marooned in a sea of mud whose youngest baby, Amoth, was found alive, floating among a sea of corpses. It was, said the BBC's Andrew Harding who had voiced the report, a unique glimpse into Burma's misery. But what was unusual was that this report, lasting two minutes and 27 seconds at the top of the programme, had not been filmed by the BBC or by Harding. All but 32 seconds had been filmed by Jonathan Pearce, of the aid agency Merlin (Cooper, 2011).

It must be stressed that this was well signposted by the BBC; it made clear that the material had come from Merlin and there was a piece to camera from Jonathan Pearce during the package. And the BBC were not the only ones to rely on aid agency content.

> Often we'll use photos and quotes that have come from aid agencies. If you're stuck in the UK, your options are more limited.
>
> (Interviewee 28, broadsheet writer)

> I suspect that people rely a lot more on the NGOs than they did, and often you'll see newspapers running columns by a Save the Children doctor or what have you . . . because people don't have the resources any more, certainly with photographers, I think. So you'll have NGOs on the ground taking images that the media uses, whereas previously you'd have sent your own person on the plane. But now it's a toss-up whether [you] send one person never mind three or four.
>
> (Interviewee 25, foreign correspondent, daily broadsheet)

The fact that aid agencies were hiring former journalists meant that their work was more trusted by the broadsheets; in some cases there was even a blurring of lines about whom the journalist was actually working for (Cooper, 2007):

> I remember one time in Iraq, when someone left our newsroom to go and work for [a major aid agency], and then he rang me up from a hill in Iraq overlooking the scene of some terrible sight, and he gave an account of it. There he was, he was the [aid agency] man on the spot, but he may well have provided pictures to go with the account, and of course I would be very happy to take that because he's somebody that I trust completely, and I trust that organisation to a fair degree as well.
>
> (Interviewee 20, foreign editor, Sunday broadsheet)

The main problem was labelling. Both journalists and aid agencies admitted that there was often a lack of clarity about who had been the original author of the content. There were various reasons why journalists might not label it correctly—confusion, or lack of desire to admit the true authorship—and aid agencies did not always push for a correct credit, because for them the fact it was being used by a mainstream media organisation got the message across.

> I'm not sure that you do label it, particularly. Maybe in certain contexts you would infer it from the way it was. I think sometimes you might say "interviews conducted by", and then name the charity, but not always necessarily.
>
> (Interviewee 28)

The concept and execution of some pieces from aid agencies meant they were not only worthy of making a newspaper, but could even make the front page. The Sunday broadsheet foreign editor recalled running a splash on the food crisis in the Sahel region of Africa. A major aid agency had approached the paper with photographs and written case studies of women who were binding their stomachs with rope to try to ease their hunger pains; the piece itself was then written up by the foreign editor and a freelance journalist.

> It is an absolute classic case of cooperation between an aid agency and a newspaper, not on a big breaking story but they knew we were interested in the issue and they came with us a very good, if you like, crude journalistic term I know, 'angle', on a developing bad situation. . . . And I see nothing wrong in that relationship; they're not a commercial company, they're not doing bad things, they're only doing good. They're a very good outfit, they're a member of the DEC. I wouldn't probably deal with an agency that was not a member of the DEC and I'd certainly have a certain amount of squeamishness about proselytising religious aid agencies.
>
> (Interviewee 27: foreign editor, Sunday broadsheet)

The foreign editor of a daily broadsheet (Interviewee 19) took the view that he would never use words from an aid agency (he believed that should be confined to first-person pieces on the op-ed pages), but that pictures were sometimes acceptable. This was the attitude of another foreign editor for a daily broadsheet (Interviewee 24), who thought the 'by-line' in the news pages was sacrosanct:

> I think it's clear: if it's presented as the objective voice of the paper then it should not be them [aid agencies]. If it's presented as someone else's voice with a picture and a by-line at the end then that's totally fine with me. But if you were using them in a news piece then it should be quite clear that's who they are. I suppose, to be scrupulous about it, you would certainly have unattributed reporting in a story if it was something that was reliable and could feed into that stuff.

The editor of the flagship news programme, saw aid agency footage as "the bane of his life". His objections were that the quality was not good enough; it rarely told a story in journalistic terms; it lacked focus or was the aid agency's story rather than the story of the people there. He felt his programme was particularly targeted by aid agencies because they saw it as more sympathetic to international stories (Interviewee 35). An international editor for a commercial broadcaster said his employers would be extremely reluctant to use aid agency footage, but that there were circumstances in which he could imagine it happening.

> To be absolutely honest, if we can't get our cameraman in to film something, then there's something wrong. But if they have an extraordinary video from an area where they've been and no-one else has access to it, I don't see a problem with using that as long as you very carefully say where it came from, when it was shot—"this is a Save the Children video and Save the Children are saying there is no government help in this area". And then, of course, you say we've contacted the Sudanese government and they say they've tried very hard or whatever. You need a counterbalancing statement. You do need to label up where your content is coming from.
>
> (Interviewee 38)

His view was echoed by his editor and by the head of international news at another broadcast organisation, who took a more relaxed view about the use of aid agency video (Interviewee 39). The foreign editor said that, in a "utopia", their organisation would like to have shot everything themselves, but if the footage was of somewhere they couldn't get to and the aid agency could then they would use it. He saw a new relationship emerging, where it would be more useful for the NGOs to put their footage directly onto the web without the 'barrier' of a broadcaster.

> NGOs don't need material to be laundered through a broadcaster anymore. So in a way the internet's quite liberating to them, and allows us to streamline our relationship with NGOs in a sense, and actually just focus on what has the maximum value. Because I think in the past stuff's been pumped out from NGOs' press operations and a lot of it never sees the light of day, and I think NGOs can now be far more targeted at what they're doing—their own web operation, then logistical help to broadcasters to get to places.
>
> (Interviewee 37)

For the aid agencies, there were different attitudes to what they wanted in return for their content. For the smaller agencies, any interest from the media was worthwhile.

> People will come to us for different things. Sometimes people come to us for access, or background, so you don't want any credit for that one. There are a few journalists who will ring us up if they're going to Mali, and say "Save the Children can't get us into this town, can you get me into that town with an interpreter, or can we speak to your guy in Bamako while we're there?" You know it's not a proper interview, it's just for some "how's it going, and what does it really feel like?" I think that's really important, because if they weren't ringing us for that, we would just have no chance.
>
> (Interviewee 17, head of media and corporate communications, NGO)

For the larger, more established agencies there was concern about getting the credit, particularly when film footage had been used. The head of media at one aid agency gave an example of what had happened to him at a previous organisation:

> I have had experience in the past of providing the BBC in particular with footage, and when it was broadcast they had gone out of their way to remove the name of the organisation which I was working for, which had instigated the coverage. It was so mealy-mouthed and it was so bloody-minded on their part that it left a bad taste in my mouth.
>
> (Interviewee 8)

Aid agencies' main problem, though, was getting the footage used in the first place. They reported media outlets resisting the use of their video news releases (VNR). There was one notable exception to this tendency: when the footage involved a celebrity. One small agency claimed journalists rarely used their footage normally, but when they took Hugh Quarshie, a *Holby City* actor, out to north-east Kenya during the East Africa drought of 2011, there was a huge pick up. But for agencies like them, any credit was seen as a bonus; although often the filming went uncredited while spokespeople were given astons[9] clearly identifying them as a member of agency staff.

But there were times when the agency—despite having facilitated a trip and perhaps even paid for it—would request that their name was not used in any way. This was usually because the journalist planned to write something critical of the government in question, which might endanger aid workers in the field—a problem one agency had in Zimbabwe and South Sudan for example. Such concerns might even make aid agencies ask to look at the copy of less experienced regional reporters.

> I've always made clear, if we take regional newspapers and pay their fares, we will ask them to show us their copy so that we can check it for facts, factual errors. I actually don't expect the regional press to get particularly opinionated about it. With the national media, we're not in a position to control it. What we can do is we can say "please take [our]

name out", but that's all. I wouldn't dream of even trying to determine what they should write or what they shouldn't.

(Interviewee 8, Head of Media, INGO)

Conclusion

When reporting humanitarian disasters, journalists often come to depend on aid agencies for help in a way they would reject, or look for more diverse sources, in other situations (Ryle, 2000; Cooper, 2007). News desks defend this on the purely practical basis of needing to get the story out. They point out that the journalist's and media outlet's commitment to professionalism and objectivity mean that there should be few concerns about these close—or symbiotic—relationships.

However, journalists tended to portray aid agencies as the 'good guys' in humanitarian stories, in a way that they wouldn't with commercial organisations. Even those reporters who had fairly critical remarks to make about aid agencies typically prefaced their comments with a disclaimer:

> I must say, I became quite, this is absolutely not to talk in any negative way about the primary focus of their work. . .
>
> (Interviewee 24: foreign editor, daily broadsheet)

> I don't want to have a pop at the aid agencies, they're marvellous, they do a great job and all the rest of it.
>
> (Interviewee 40, foreign affairs editor, satellite broadcaster)

Journalists typically rejected the view that they were significantly dependent on aid agencies, although the most hard-up newspapers did admit to using agency resources frequently, and the NGOs themselves reported far more often than journalists that they had set up fixers and translators, and paid for cars and accommodation. In more recent times—as aid agency offices have become more professional, and access to high-quality photographs, videos and text has increased (Cooper, 2011; Wright, 2015)—journalists also appear more willing to use content produced by aid agencies as the basis of a report. As such, the relationship between the two becomes a mutually beneficial one, which is often ongoing and recurring.

Notes

1. https://twitter.com/jamesrbuk/status/203873386030563328
2. In 2011, Save the Children formed what was called an "ambitious and strategic global partnership" with pharma giant GlaxoSmithKline in which GSK promised to reinvest 20 percent of the profits it makes in the world's least developed countries back into projects to strengthen healthcare infrastructure (www.savethechildren.org.uk/about-us/who-we-work-with/corporate-partnerships/our-partners/gsk). Justin Forsyth, then CEO of Save the Children UK, described the partnership with GSK as follows: "Many years ago I used to campaign against [GSK] because they didn't do enough on HIV and AIDS drugs. I used to picket their annual

meeting years ago; now they are at the forefront of investing in a malaria vaccine, partnering to invest in health workers. We campaigned together last year" (interview, London, 17 January 2012).
3. For example, the *Independent* newspaper launched an appeal following the East Africa food crisis of 2011 after a journalist was sent to a refugee camp (see independent.co.uk/giveadayspay) and then also launched their annual Christmas appeal with the iNGO Merlin, who were working there.
4. See Alagiah (1998) for example.
5. Cottle and Nolan are comparing the journalists' experience with those embedded with army divisions during the Gulf Wars.
6. Hamer and Coburn were attacked while embedded with the US Marine Corps in Afghanistan on 9 January 2010. Their vehicle hit an improvised explosive device (IED) near Nawa in Helmand province. Hamer was killed and Coburn suffered severe leg injuries.
7. Interviewee 17 is referring to paying a day rate/fee for a journalist to cover the story.
8. Interviewee 6 was clear however that they expected journalists to pay their way.
9. On-screen text naming the person filmed and usually their job title or company.

References

Abbott, K. 2009, 9 November-last update, *Working together NGOs and journalists can create stronger international reporting* [Homepage of Nieman Lab], [Online]. Available: www.niemanlab.org/2009/11/kimberly-abbott-working-together-ngos-and-journalists-can-create-stronger-international-reporting/ [2015, 18 August].

Alagiah, G. 1998, 27–28 May, *Dispatches from disaster zones: Reporting humanitarian disasters*, Conference, British Red Cross, London.

Altheide, D.L. & Snow, R.P. 1979, *Media logic*, Sage Publications, Beverly Hills, CA.

Brook, S. 2008, 3 September-last update, *Jobs to go in daily mirror cutbacks'* [Homepage of The Guardian], [Online]. Available: www.guardian.co.uk/media/2008/sep/03/mirror.pressandpublishing [2015, 15 August].

Cooper, G. 2007, *Anyone here survived a wave, speak English and got a mobile? Aid agencies, the media and reporting disasters since the tsunami*, Nuffield College, Oxford.

Cooper, G. 2011, *From their own correspondent? New media and the changes in disaster coverage: Lessons to be learned*, Reuters Institute for the Study of Journalism, Oxford.

Cooper, G. & Cottle, S. 2015, "Humanitarianism, communications and change: Final reflections", in *Humanitarianism, communications and change*, eds. S. Cottle & G. Cooper, Peter Lang, New York, pp. 251–265.

Cottle, S. & Nolan, D. 2007, "Everyone was dying for footage", *Journalism Studies*, vol. 8, no. 6, pp. 862–878.

Cottle, S. & Nolan, D. 2009, 16 November-last update, *How the media's codes and rules influence the way NGOs work* [Homepage of Nieman Journalism Lab], [Online]. Available: www.niemanlab.org/2009/11/simon-cottle-and-david-nolan-how-the-medias-codes-and-rules-influence-the-ways-ngos-work/ [2015, 18 August].

Deacon, D. 2003, "Non-governmental organizations and the media", in *News, public relations and power*, ed. S. Cottle, Sage Publications, London, pp. 99–114.

Edwards, A. 2013, 6 March-last update, *Number of Syrian refugees reaches 1 million mark* [Homepage of UNHCR] [Online]. Available: http://www.unhcr.org/uk/news/latest/2013/3/513625ed6/number-syrian-refugees-reaches-1-million-mark.html [2018, 23 August]

Fenton, N. 2010, *New media, old news: Journalism & democracy in the digital age*, Sage Publications, Thousand Oaks, CA, London.

Hamilton, J.M. & Jenner, E. 2004, "Redefining foreign correspondence", *Journalism: Theory, Practice & Criticism*, vol. 5, no. 3, pp. 301–321.

Hannerz, U. 2012, *Foreign news: Exploring the world of foreign correspondents*, University of Chicago Press, Chicago.

Harding, P. 2009, *The great global switch-off: International coverage in UK public service broadcasting*, POLIS, London School of Economics and Political Science, London.

Manheim, J.B. 1998, "The news shapers: Strategic communication as a third force in news making", in *The politics of news: The news of politics*, eds. D. Graber, D. McQuail & P. Norris, Congressional Quarterly Press, Washington, DC, pp. 94–109.

Moore, M. & Loyn, D. 2010, November-last update, *Shrinking world: The decline of international reporting in the British press* [Homepage of Media Standards Trust], [Online]. Available: http://mediastandardstrust.org/wp-content/uploads/downloads/2010/11/Shrinking-World-FINAL-VERSION.pdf [2015, 17 June].

Murrell, C. 2010, "Baghdad bureaux: An exploration of the interconnected world of fixers and correspondents at the BBC and CNN", *Media, War & Conflict*, vol. 3, no. 2, pp. 125–137.

Murrell, C. 2013, "International fixers: Cultural interpreters or 'People Like Us'?", *Ethical Space*, vol. 10, no. 2/3, pp. 72–79.

Murrell, C. 2014, *Foreign correspondents and international newsgathering: The role of fixers*, Routledge, New York.

Noor, I. 2005, 24 October-last update, *Doctor's diary: Healthcare without hospitals* [Homepage of BBC], [Online]. Available: http://news.bbc.co.uk/1/hi/world/south_asia/4347374.stm [2011, 6 November].

Palmer, J. & Fontan, V. 2007, "'Our ears and our eyes' journalists and fixers in Iraq", *Journalism*, vol. 8, no. 1, pp. 5–24.

Powers, M. 2015, "NGOs as journalistic entities", in *Boundaries of journalism: Professionalism, practices and participation*, eds. M. Carlson & S.C. Lewis, Routledge, Abingdon and New York.

Ryle, J. 2000, "The hazards of reporting complex problems in Africa", *Transnational Law and Contemporary Problems*, vol. 10, pp. 85–106.

Waisbord, S. 2011, "Can NGOs change the news?" *International Journal of Communication*, vol. 5, pp. 142–165.

Wright, K. 2014, *A quiet revolution: The moral economies shaping journalists' use of NGO-provided multimedia in mainstream news about Africa*, PhD submitted to Goldsmiths University of London.

Wright, K. 2016, "These Grey Areas' how and why freelance work blurs INGOs and news organizations", *Journalism Studies*, vol. 17, no. 8, pp. 1–21.

Wright, K. 2018, *Who's reporting Africa now? Non-governmental organisations, journalists and multimedia*, Peter Lang, New York.

4 The "First Draft of History" or Just "Smoke and Mirrors"? How Journalists and Aid Agencies See the Use of UGC in Humanitarian Crises

> Twitter's the very, very first draft of history now. It tells you things that are happening before most other traditional ways of airing news tell you. So if something's happened it will be on Twitter half an hour, 45 minutes before it's on any other media form.
>
> (Interviewee 21 foreign editor, broadsheet)

> Death toll bingo . . . has got a lot worse. The tendency of editors to say "Sky says 5,000 dead now, someone just tweeted there's 7,000 dead" and they don't know anything about the situation. We all know the riot in Strangeways Prison, where scores of prisoners were said to have been impaled; it turned out [there were] two dead. The famous Coach H, Paddington rail disaster, hundreds incinerated, not true. Hurricane Katrina; ten thousand dead inside the convention centre, not true.
>
> (Interviewee 27, foreign editor Sunday broadsheet)

The Indian Ocean tsunami of 26 December 2004 changed the way major news stories were reported forever. It was not the first event using user-generated content (UGC), but it was the first disaster where the dominant images came from ordinary people.

So coverage of the tsunami threatened to challenge the very nature of journalism itself—and the role it prided itself on of bearing witness and breaking and distributing news. News stories used to be journalist-only, boundaried spaces, where ordinary people only appeared as archetypes in narratives or sources for comments such as a case study. The tsunami changed this; perhaps the most vivid exposition of this was when journalists and producers were sent to airports, not to get on flights to the tsunami zone but to pick up holidaymakers' video at the airport (Burrell, 2005).

This chapter asks those have who traditionally acted as fact custodians and truth tellers when it comes to humanitarian disasters about how their practice has been altered by user-generated content. It uses interviews from editors, reporters and aid agency press officers to analyse how user-generated content is perceived and used by these three different but interlinked groups. Editors, back at their desk and with responsibility for final content, may perceive the use of UGC very differently to their reporters out covering a story. Aid agency press officers, used to dealing with the mainstream media, are facing new challenges and opportunities in the way social media and UGC are bringing humanitarian crises to the fore.

The first section begins by asking editors and reporters about their reactions to the use of UGC during humanitarian disasters, while the second section asks them to discuss ethical questions in practical terms. The third section turns to aid agencies to ask how they have been challenged by the growth of UGC, and how they, too, are using it to ensure a disaster is put and kept in the public eye.

For the aid agencies interviewed, user-generated content via social media also had a potentially profound effect. They were no longer the exclusive gatekeepers of messages coming out of disaster zones. This altered their relationship with both reporters and the beneficiaries they had previously spoken for. So it was not just about losing control of the message, but also the implications this had for fundraising and advocacy.

User-Generated Content and Challenges to Norms of Journalism

There are several key areas where user-generated content is acting as a challenge to the established norms and practices of journalism, transforming ways of storytelling and truth claims. Overarching all this is its challenge to what journalism actually is, and how porous its boundaries are becoming (Carlson and Lewis, 2015).

Bearing witness—'being there'—has long been held as one of the key characteristics of journalism and one of its justifications. But the proliferation of communication technologies such as mobile phones, the internet and social media has meant that the eyewitness role is no longer confined to the 'professional' journalist working for a mainstream media organisation. Often, in major stories, journalists themselves did not witness a crisis incident but were first on the scene in its aftermath to report back to a wider audience. Now that privileged role of producers of truth and information is being challenged by those who were actually caught up in a quake, a flood, a landslide and can take pictures or send text to report on it.

So when someone like a security guard captures the Moore tornado on their phone, as detailed in Chapter 7, it brings into question the role of bearing witness—and beyond that even the very idea of who is a journalist. Could his six-second videos be considered a form of journalism? His work was certainly used by a variety of mainstream media until, feeling stressed and under pressure, he withdrew it.

Journalism, with its fraught relationship with professionalism, particularly in the UK[1] has always been inherently fluid—Bourdieu describes it as a weakly autonomous field. But journalists themselves view their work as important and "defend it against incursions from non-journalists" (Carlson,

2015:9). The result has been, during this time of rapid technological change, that the journalistic space has become increasingly contested, with boundaries questions. Carlson adapts Gieryn's work around science to explain this. Gieryn (1999) sees actors struggling to lay claim to credibility—in this example, journalism by either expulsion (struggles around who is authoritative), expansion (struggling to expand journalism into other spheres) or protection of autonomy (in which 'experts'—in this case journalists—try to erect interpretive walls to prevent 'non-journalists' co-opting their authority). Gieryn's ideas dovetail with Bourdieu's concept of the struggles by competing agents around the boundaries of the journalistic field (as detailed in Chapter 2). As the interviews with editors and reporters make clear in this chapter, journalists frequently questioned the value and use of images and text created by members of the public caught up in humanitarian disasters and its value compared to their own output. The result is a demarcation between 'insiders' and 'outsiders' with claims over expertise frequently expressed in terms of what *is* or *is not* journalism.

The wider field of journalistic ethics, circumscribed by the kind of laws, codes and norms of practice outlined in Chapter 8 are also challenged by the accession of these new agents into the journalistic field. 'Accidental' journalists—the kind of UGC creators whom I interviewed—did not define themselves as journalists and as such did not see themselves held to account by journalistic codes. This was not to say that they did not define an ethical framework for themselves but this was mainly articulated around a bearing witness to what they saw as truth—not dissimilar to Jeffery P Jones's observation of 'truthiness' which he saw as a move from "a journalist-centred regime of truth to a place where citizens, politicians, various institutions and groups perform creative constructions of truth"—(Tenenboim-Weinblatt, 2009:102).

The way the UGC creators most frequently distinguished their moral centre though was by articulating their refusal to make any (monetary) profit out of other people's suffering—that is they made no charge for the images that they shared via social media which were then taken up by the mainstream media.

This ability of the public to publish and share within seconds has also given rise to questions over the construction of meanings around crisis stories. In the aftermath of the south Asian tsunami of 2004 there were claims that the explosion of user-generated content would mean that creators of content would be be active shapers rather than passive bystanders in their own stories (Gillmor, 2005; Glocer, 2006; Jenkins, 2006; Deuze, 2007). As a result, disasters today are not just something that audiences 'see' in the media but something they are "increasingly socialised to create" (Frosh and Pinchevski, 2009).

The result is often raw, immediate and subjective images and text, a move away from the "anaesthetizing effects" of traditional disaster reporting (Chouliaraki, 2006:25) acting to minimised distancing effects (Chouliaraki, 2008).

Conventional understandings of journalistic ethics frequently see objectivity as the foundation of the journalistic profession (Tuchman, 1972; Gans, 1980; Schiller, 1979; Schudson, 2001; Schudson and Anderson, 2009), with the journalist as a central, impartial, distanced witness to events. In coverage of disasters, this has been challenged by those who subscribed to the idea of cosmopolitan sensibility (Chouliaraki, 2006, 2008; Boltanski, 2000; Seaton, 2005).

Audience focus groups have consistently responded positively to such emotional coverage with journalists also recognising its power (Wahl-Jorgensen, Williams and Wardle, 2010; Williams, Wardle and Wahl-Jorgensen, 2011). As a result, journalists have sought to normalise this and co-opt the strengths of UGC by adopting a more personalised, emotional tone in their reporting.

The contestation over objectivity has also led to an increasing importance of transparency. With bloggers and UGC creators publishing immediate if incomplete accounts and making clear the positions that they may adopt (such as activist), this has put pressure on journalists to also adapt to what has been called 'process' rather than 'product' journalism (Jarvis, 2009) by amending, updating and being transparent about their sources and approach.

The advent of UGC has also resurrected questions over the kind of imagery that is now acceptable—the idea of the citizen journalist becoming the 'citizen paparazzi' (Niles, 2005) sharing gruesome images without thought, as has been criticised in both coverage of the Bangkok blast (Brown, 2015) or Jordi Mir's images of the killing of Ahmed Merabet (Satter, 2015).

An established journalistic norm has been the so-called 'breakfast table test' (what newspaper editors believe their readers can bear to see over the breakfast table). The result is that UGC, whose creators are not signed up to such unofficial norms, has forced news organisations to make constant ethical statements about what they are willing to publish or not publish. If they use UGC which shows imagery of disaster or death, they may to choose to frame it by cropping the pictures, using black and white instead of full colour, using it only on certain parts of their bulletins, paper or website or prefacing it with some kind of warning about the kind of content to follow: they are implicitly and explicitly referring to their editorial judgements on what is 'acceptable' or not.

The result is also that verification of UGC has become a part of the journalists' job (Singer, 2014), as has the concept of gatewatching (Bruns, 2008), with many journalists feeling unease that their job now includes not only conventional reporting but curation and aggregation—seeking out such content, and ensuring they beat the opposition by finding it first.

The Journalists' View of User-Generated Content

It was no surprise that sharing user-generated content via social media was now an accepted part of journalistic practice, particularly in the light of budget cuts and staff layoffs at a time of a 24-hour news culture. This did not mean, however, that it was accepted or welcomed into the field of journalism. Traditionally, mainstream media journalists tell stories as a way to establish a certain value system, transforming the "raw stuff of experience into presumed fact and arrange facts into coherent stories" (Jamieson and Waldman, 2004:xvii). User-generated content—with its diffuseness, subjectivity and lack of coherence—was often seen as a distraction. Far from being a replacement for journalism, it was seen as demonstrating the need for 'trained' and 'professional' journalists to moderate it. What was interesting, however, was that, while one might expect desk-bound editors (with a need to get as much content as possible) to take a different view from reporters on the ground, there was often confusion and disagreement about the value of UGC in disaster reporting on both sides.

The Editors' View: "Whatever We Can Lay Our Hands on"

Bruno (2011) coined the phrase the '1440-minute' cycle, to reflect the sped-up nature of the news cycle in a technologically advanced age. The need to rapidly disseminate information in an increasingly competitive news environment was what attracted many editors to user-generated content, particularly if it was shared via social media.

With newspapers constantly updating their websites rather than waiting to go to press and TV news either having an online or 24-hour broadcasting presence, the need for fresh content has become paramount, as well as the fear of being left behind (Hermida and Thurman, 2008). For the midmarket foreign editor, it didn't matter who the Facebooker/Tweeter was, as long as they were actually there, "though if they are in a position of authority that is better for verification" (Interviewee 32). The assistant editor (newsgathering) at a commercial broadcast organisation pointed out that the sheer amount of time it still took to get a correspondent to a disaster zone meant that, on the first day in particular, the coverage would be largely driven by sources that were not his own journalists:

> Picture an earthquake in India. We're flying in from Dubai, our nearest bureau; the flight leaves at 4 in the afternoon UK time. Obviously they're in the air when the 6.30 [bulletin] comes on. So then we'll assemble that piece through a combination of agency pictures, our local stringer on the ground, tweets, any picture that's posted on the web, aid agencies, we might do a phone interview or Skype with someone from Oxfam who happens to be in the affected region, or Save the Children or whoever. So in those sorts of circumstances, it's whatever we can lay our hands on, really.
>
> (Interviewee 37)

Those editors who were positive about UGC saw more explorative and dynamic ways of using such content, and talked passionately about how it could be a valuable tool. One broadsheet foreign editor described how, after the Haiti earthquake, his paper had produced a timeline graphic of the quake using tweets to illustrate and punctuate it (Interviewee 34). The midmarket foreign editor saw it as part of a continuum: that pictures had always been a vital part of humanitarian stories and that UGC images were a way of sending a story up the news agenda—and could on occasion even result in overload.

> I think disasters are now covered more extensively because of the internet and accessibility. If you look at the Nepal earthquake, the coverage was enormous. In fact, I think there is now so much overkill with reporting that the problems are not alerting the world to what has happened in a remote part of the world, but rather not turning them off from a 'story'.
>
> (Interviewee 32)

The editor of a flagship news programme saw UGC as a way of leading journalists to previously overlooked parts of the story while a broadsheet foreign editor saw it as indicative of the opening up of society—contrasting the way that the explosion of social media in China had made suppressing events extremely difficult compared, say, to what was still possible in North Korea or Burma (Interviewees 35, 21).

Yet many editors still verbalised a considerable degree of scepticism about UGC. They stressed the importance of having a traditional journalistic presence 'on the ground', rather than being confined to the role of an aggregator or curator of content. Many interviewees emphasised the value of on-the-ground journalism even though they went on to complain about the lack of budget and staff, thanks to well-documented, widespread cost-cutting and reductions in foreign news bureau correspondents, staff and travel budgets (Harding, 2010; Sambrook, 2010).

Despite the increasing use of Twitter to construct news stories,[2] editors often denigrated UGC as nothing more than a temporary stopgap to appease a voracious 24-hour news machine. The Sunday broadsheet foreign editor saw Twitter as largely irrelevant for his reporters—other than for last minute 'mopping up' of information on Saturdays—but he did see videos that they could source from social media as "a very attractive way of drawing people in" (Interviewee 20).

Another broadsheet foreign editor, and one of the most experienced journalists interviewed, was the most scathing when it came to the increasing use of UGC social media by journalists. For him, the personal and subjective was irrelevant, and having to sort through thousands of tweets was impractical in a breaking news situation.

> I mean, citizen journalism is all very well but when it comes to [a disaster] happening at 3 o'clock on a Saturday afternoon, I've gotta write 1,500 words by 6 o'clock, we haven't got time to wade through every bloody tweet on a Saturday because invariably they're all "oh my heart goes out, my thoughts are with . . ." Who the fuck cares? Big bloody deal. So [social media's] clogged up with just unutterable crap and gesturing and emoting all over the place.
>
> (Interviewee 27)

Even those in favour of UGC saw its limitations—for example, the digital divide meant that some crises would get more coverage than others[3]—but for most it was unimaginable to construct a story without some form of UGC.

The Reporters' View: "Smoke and Mirrors"

Editors and reporters perform different roles, so it was important to try to tease out the differences in their attitudes to UGC. The image of the reporter on the ground, using social media to communicate and gather user-generated content, was not borne out in my interviews. Many of the reporters I spoke to were veteran foreign correspondents who had worked in a pre-social-media age, which might account for some of the resistance expressed to the use of UGC. But, even for the younger ones, UGC and social media were seen largely as a distraction that interfered with the practicalities of reporting to deadline.

As one broadsheet foreign correspondent pointed out, power outages meant that it was often difficult to access UGC, and in the scramble to get to the story there was no time to monitor social media:

> The trouble was [during Typhoon Haiyan in the Philippines], there was no electricity and no real internet at first, and a kind of haphazard phone, no data signal. I remember writing something about Imelda Marcos; it took me all day from waking up, trying to log on to find out about her shoes. It wasn't something where I really had the equipment for using social media. I wrote a story *about* social media, with a doctor who raised funds on social media and set up an emergency mission, but reporting it, I didn't [use social media], no.
>
> (Interviewee 25)

Arguments over what is or is not journalism are often struggles over boundaries. They are often symbolic contests in which actors vie to control definitions over what is or is not labelled journalism (Carlson, 2015:2). The reporters commonly chose to define UGC as non-journalism by taking a 'top down' view of the content. They classified it as nothing more than source material for interviewees, facts or 'colour'. The broadsheet feature writer summed it up as follows: "Tweets for instant reaction . . . Facebook for trying to get hold of somebody to send those grim messages: 'Hi . . .' " (Interviewee 28).[4]

The international editor at a commercial broadcaster[5] said it was now standard practice to ask in the middle of breaking stories on disasters whether anyone had taken footage or stills on their mobile phone. But both he and a Sunday broadsheet foreign correspondent did not articulate that as a use of user-generated content, rather just a technologically sophisticated way of gathering what was previously known as the 'collect'[6] (Interviewee 40, 22).

In fact, a veteran foreign correspondent for a broadsheet was the only reporter who stressed the importance of UGC in humanitarian crises. He overtly drew on former prime minister Gordon Brown's claim that the genocide in Rwanda would not be able to happen again now the world had Twitter (Viner, 2009).

> I've just been in Rwanda, actually, and one of the things going back to the town I was in during the genocide that was most striking was that 20 years ago this place felt completely isolated, in the middle of nowhere. You could see how they could murder 11,000 people in a single day in a church and nobody in the rest of the world knew. Now they've got cellphones and all the rest. One of the young women I wrote about, within two days she'd emailed me to say she'd seen the piece and could I send her the pictures. And therein lies the difference.
>
> (Interviewee 30)

But his attitude was unusual. Reporters saw user-generated content as a direct challenge to their normative standards of journalism and would often use pejorative terms to describe it. Interviewee 22 dismissed it as "smoke and mirrors", while an international editor at a flagship news programme felt that journalists and editors were often using UGC without fully recognising the ethical problems it could bring—particularly the inherent biases, promotion of causes and inability to tell the full story. While reporters framed this concern as an issue about standards of journalism, it often segued into worries about their own job security; as she described it, why would any editor spend money sending a young journalist to a crisis zone "when they're getting perfectly good wobbly pictures from some person in the middle of it" (Interviewee 36).

Journalists and the Ethical Issues Around UGC

However reluctant editors and reporters were to acknowledge UGC (particularly gathered via social media) as an alternative form of storytelling or truth-gathering, it had made practical differences to the way they carried out their jobs.

Verification

An analysis of the top 100 most-tweeted picture stories for the *Guardian* datablog showed that 15 percent of the 2012 Hurricane Sandy pictures were fakes (Burgess, Vis and Bruns, 2012). Following the death of the Al-Qaeda leader Osama Bin Laden in May 2011, pictures that claimed to show his dead and mutilated body were distributed on social media and rapidly picked up by the mainstream media (Newman, 2011). Meanwhile in the aftermath of Hurricane Harvey in 2017, the online fact-checking organisation Snopes reported that a viral video posted on Facebook claimed that the Red Cross were stealing donations from churches in Texas and then selling or burning them (Snopes, 2017).

The need to verify UGC has become increasingly important for mainstream media organisations, but it quickly emerged that there was no single standard for verification, and often huge amounts of confusion. As Shapiro et al put it: “Verification for a journalist is a rather different animal from verification in scientific method, which would hold every piece of data subject to a consistent standard of observation and replication” (Shapiro et al, 2013:668–669).

Yet coverage of the 2011 ‘Arab Spring’ uprisings and of the Syrian conflict often relied heavily on UGC, because journalists had limited access to the areas where the news was happening. Activists learned how to manipulate the story for their own benefit; a 2012 report by *Channel 4 News* from Homs about the video activist Omar Talawi revealed how the activists embellished a report by burning a tyre behind them (*Channel 4 News*, 2012). It was not until 2014 that a handbook, authored by a mixture of academics and journalists, suggesting how to verify digital content in emergencies was published (Silverman, 2014).

The Verification Handbook has a checklist detailing no fewer than 42 online verification tools (2014:108–110), which can be used to confirm the identities of content creators, where content was created and the authenticity of images. But when I asked the journalists about how they verified UGC, there was still a strong sense of relying on ‘gut instinct’, linked to journalistic experience, in assessing whether something was fake or genuine (and whether other people were using it). As a Sunday broadsheet foreign editor put it:

> I think, frankly, verification is like, ‘does it smell right?’ ‘Are the BBC doing it?’ There’s quite a lot of reliance on ‘are reliable people using it?’, which is not a good way to proceed.
>
> (Interviewee 20)

When asked specifically about the kind of technical checks that could be run on UGC, very few interviewees had knowledge of geopositioning, mapping technology or even the concept that verification is a process rather than the idea that something can be instantly identified as right/wrong (Wardle, Dubberley and Brown, 2014). The foreign editor at a commercial broadcaster was the only journalist outside specialist areas[7] able to give a detailed answer about verifying UGC—mentioning checking seasons, weather, location, clothes, dates and signposts, as well as prioritising getting hold of its creator (Interviewee 39).

UGC was seen mainly as an issue for 'desk' journalists, rather than reporters on the ground. The exception was when a correspondent sourced exclusive content themselves, such as Interviewee 36's sourcing video of jihadis seeking retribution in Gao, Mali, in 2013. She said she felt confident in verifying the UGC because she had been introduced to the creator by someone she trusted; the creator had been able to answer questions about where the footage was taken and when, and who was involved.

But more reporters understood that verification was becoming part of the journalistic 'skill set'. For many there was often a sense of vulnerability and unease. One reporter talked about being in the middle of a story, having a mobile phone thrust into his hand with a supposed film of an atrocity, and being very conscious that "you're in a country and a culture that is alien to you and you don't speak the language, there are things that might stand out to a native speaker, that something doesn't stand out to you" (Interviewee 22). There was also a fear that editors in London would encourage them to use UGC, despite the fact that the reporters themselves had had no role in its verification.

> If I'm in a place and I'm rung by the desk and they say, "We've got this amazing piece of footage, which I think you should consider including in the package, it shows some x, y and z," then I'm in their hands, and you just trust your colleagues. We caveat it with those horrible phrases, "we can't guarantee the authenticity of this, we've got no proof that this happened, a certain group claims that this shows x and y". You can distance yourself somewhat from it so at least you're not saying I know this to be true, but it's difficult, you know.
>
> (Interviewee 30, international editor, commercial broadcaster)

Very few journalists spontaneously mentioned the use of verification agencies such as Storyful. Many relied on news agencies such as Reuters, AP and AFP to run checks if UGC was transmitted on the newswires. Interestingly, apart from the *Guardian*, which had attempted to use crowdsourcing to interrogate stories such as the MPs' expenses scandal (GNM press office, 2009), a foreign editor from a broadsheet was the only person to vocalise the idea of the audience being part of the verification process:

> If we think it's sufficiently important, we could say this is unverified, we think this is the area, this is where it's come from, lay the facts out

> for people and they can decide for themselves, or you can crowdsource it, say "where does it come from, anyone know, anyone recognise that building?"
>
> (Interviewee 21)

The idea of being hoaxed was one that many journalists feared, although they were reluctant to admit it ever happened to them; most of the stories they recounted were about other newspapers' problems or 'near-misses' where a lucky call had saved them from putting up an erroneous story.[8] So the BBC's then social media editor Chris Hamilton's candid blogpost, about being hoaxed over a photograph purporting to be of the Houla massacres in May 2012, was a rare example of a media organisation admitting it had got it wrong and describing its attempt to deal with its mistake (Hamilton, 2012); this honesty, though, was in the aftermath of extensive media criticism from around the world (Furness, 2012; Silverman, 2012).[9]

Worryingly, there was little concern about verification because it was believed humanitarian disasters could not be 'faked'. As a foreign editor for a commercial broadcaster put it:

> I don't think in my heart of hearts that pictures of an earthquake happening could be faked; it's quite hard to fabricate that. I'm not sure much verification would go on to YouTube footage of an earthquake, it'd be pretty much taken on trust, and I think if the industry was honest with itself, it would agree: Sky News, BBC News 24, the first pictures of earthquake damage appear in the internet tweeted, they're going to be smacked up on screen.
>
> (Interviewee 37)

There was often lack of understanding that even if quakes or hurricanes couldn't be easily 'faked' there was still a need to establish whether the pictures came from the disaster in question. For example, the *Guardian* had to take down footage from its Great East Japan earthquake liveblog because it had erroneously posted video that came from Christchurch's Cathedral Square during the 2010 earthquake there.[10]

Some of the widely shared Hurricane Sandy fakes (Burgess, Vis and Bruns, 2012) were not just the obvious ones (sharks swimming around New Jersey), but genuine pictures which had been taken on a different day—in particular, one from Arlington National Cemetery, which showed the Old Guard standing at the Tomb of the Unknown Soldier despite the appalling weather. This was posted by NPR, the *Washington Post* and the Daily Beast among others, before it was debunked by the military (Halverson, 2012; Laird, 2012).

With Sandy, the fake content went beyond just pictures. The influential tweeter @comfortablysmug (later exposed as Shashank Tripathi) started to tweet false information, such as that the New York Stock Exchange was under water—a statement that was retweeted 650 times and then made it on to the Weather Channel and CNN (Moore, 2012)—and that there would be a general power outage in Manhattan by the electricity company ConEdison. Tripathi was outed by the news website Buzzfeed and forced to apologise (Ngak, 2012). As Moore puts it:

> If Tripathi's silly tweets made it into the national press, it is the national press that is, at heart, to blame for not protecting journalistic standards as well as they should. It is a matter of a few minutes to call a spokesperson or check a live camera, and that is what journalists get paid to do. Producers or editors should not rush information to air or print until those calls have been made, and answered.
>
> (Moore, 2012)

But the pressure to get to air was one that several interviewees mentioned—and if other reputable news organisations were using the content, then it was hard to resist. Bruno's phrase "tweet first, verify later" was still apposite several years after he coined it; workload and time pressures have led to what has been dubbed "high-speed journalism" (Hofstetter and Schönhagen, 2014).

During disasters editors said a lot of the early reporting was based on YouTube, Twitter, Instagram and Skype messages, but that it was very much "one tiny piece" of what was going on. UGC, however dramatic or compelling, was not always the whole story and there was a danger this 'real' footage could be seen as such.

> The key challenge with this is facts. You can get some great footage of an earthquake, but does it tell you the true story? It shows you one snapshot from one camera at one moment, and it might provide a good shot, it might provide a bad shot, but it tells you nothing about how widespread it is, how many people are killed, whether people got warning. As with all UGC it often raises more questions than it answers, so you have to be careful. You have to be proportionate and careful and responsible with what you use.
>
> (Interviewee 37)

Some journalists felt that the drive towards 'online first' should be muted—that mainstream media could actually benefit by not rushing to put up UGC as quickly as possible, because it would help maintain its 'brand value' by avoiding errors. Journalists also thought that the value of UGC was being overstated. They felt academics whom had not been in newsrooms were giving UGC more significance than it actually had in everyday journalistic practice, and that this was proving a dangerous precedent.

Privacy

There have been numerous questions about how content from social media is used by journalists. Newton and Duncan (2012) and Duncan and Newton (2014) found that, while bereaved or shocked relatives often want their families' stories to be told by the media and that journalists often are well aware of the need for informed consent from interviewees in traumatic situations, there was less knowledge or care when it came to the use of unauthorised social media material, which the researchers discovered was seen as "stealing their [family member's] identity"; sometimes material was used that bereaved relatives were unaware even existed. How those who had created the content felt about its use will be discussed further in Chapter 8.

This was not, however, the issue that primarily taxed journalists. They rarely raised privacy questions over the use of UGC. There was a strong sense that by putting content in the public domain, some elements of privacy had been waived (Whittle and Cooper, 2009; Cooper, 2012).

There was an unofficial 'hierarchy', however, of which sites it was seen as acceptable to take content from. Twitter and YouTube were regarded as firmly in the public domain because of the default public status of tweets, for example; but attitudes to Facebook remained more mixed.

One broadsheet foreign editor said there was regular 'ripping off' of content from Facebook and other social media sites. Even 'ethical' media groups would succumb once they saw other news sites using such content.

> Tabloids and less reputable publications will just use everything from Facebook and not think twice about it, whereas people like the *Guardian* and the BBC will wait until everybody else is using it, and then feel like "the horse has bolted, we'd better use it as well", and I think that unofficially is more or less the case, so I don't think either us or them really do have a leg to stand on there.
>
> (Interviewee 70, liveblog editor)

One broadsheet foreign editor voiced the clearest concern about this. He saw a clear difference between Twitter as a public space where there was no expectation of privacy—and indeed which had been set up as a site to share views, news and content as widely as possible—and Facebook where he felt the public did not fully understand the privacy settings.

> I think everyone is pretty clear on what the legal stuff is, but I think if someone's just died in an accident, I think mining their Facebook—you know, there's this distinction between an expectation of privacy that is based on not being an important person, and then you suddenly become an important person in a way that has nothing to do with any choices that you've made and suddenly your life is exposed in that way. I don't think it's really fair to say "well they should've made their settings

> private if they were so worried about a picture of them and their ex-girlfriend popping up". So I do have a bit of a problem with that. But I think institutionally we have done it and I think we'll continue to. That's how the sausage[11] is made, I suppose, but I don't think that's great.
>
> (Interviewee 24)

For most of the journalists, however, the questions of privacy and consent became inextricably linked to permissions, copyright, taste and decency; they tended to be more concerned about beating their competitors to a particularly good photograph or video than about judging whether footage from a disaster would pass the 'breakfast table' test.

Consent and Copyright

When UGC appears in the mainstream media it is still commonly uncredited—one study found that only 16 percent of UGC on television had an online credit (Wardle, Dubberley and Brown, 2014). While most journalists are aware that copyright is held by uploaders and that permission to reproduce is more commonly sought than before,[12] there is confusion over how seriously to take this, given that content had already been put in the public domain.

In the whirlwind of breaking news, there was still a 'use now, pay/credit later' mentality among many journalists. Recalling the Vauxhall helicopter crash, the liveblogs editor said that pictures from bystanders were extensively used because they were no others available:

> We had a choice there, which was basically use it and then deal with the consequences afterwards, or don't run with it, and I particularly felt that we should use it, and when people get in touch then we would have to pay them, fine, but the pictures were so strong, and there were no wire pictures or anything. . . . In the end, what happened was that news agencies got in touch with those same people and bought up the pictures. I thought what would happen would be the people who tweeted the pictures would wade in and try to get paid and they haven't, and I think it's because—I hope this doesn't come across as patronising—but I think generally they're quite flattered that a news organisation is using their picture. It doesn't occur to them, I think, that in previous eras a professional photographer would have taken that picture and got paid for it.
>
> (Interviewee 70)

This was an area that generally taxed the broadcast media much more than print journalists because newspaper websites can directly embed social content without breaching copyright; footage and stills were more likely to be paid for than text; and papers had a smaller budget in the main. But just as many journalists thought that putting something online signed away rights to privacy, they felt putting up a video on Twitter or YouTube would override copyright.[13]

While creators of UGC were wary of talking about money (see Chapter 7), it was clear that if it was content that the journalists could not get any other way, they would increasingly be prepared to pay for an exclusive still or piece of footage. There was an increasing 'buy-up' culture among media organisations for the best UGC, but that was less about the actual content itself or who had filmed it and more because of traditional intermedia competitiveness—as one said: "I mean, there's a pressure to get hold of good footage . . . and not letting the bloody BBC get their hands on it" (Interviewee 36).

But the foreign editor for a commercial broadcast organisation pointed out that media organisations had a responsibility to deal with such UGC creators fairly:

> A lot of these people aren't very well educated, don't understand the concepts you're talking about, so you have a responsibility towards them, but at the end of the day they want this stuff to be shown. It's quite hard to go through an intricate legal discussion with someone who's just sent you some video. And everyone's different. Some people want to make money out of things, some others don't, some others want to expose the situation to the world, some want to expose human rights abuses. I think all we can do is abide by our code of conduct, our internal code, and the Ofcom code.
>
> (Interviewee 37)

Graphic Images

The UK media were careful about what they showed of (Western) casualties of the 2004 tsunami. Some of this became increasing irrelevant, however, as much more graphic images were posted to multiple non-news sites. In a piece following the Haiti earthquake, Philip Kennicott of the *Washington Post* (2010) singled out the *New York Times*, the BBC and his own paper as having used graphic and disturbing images, making the point that this was not some kind of excess confined to tabloid media.

One tabloid reporter felt there had been what she called a constant pendulum swing between "squeamishness and bloodthirstiness in news judgement" (Interviewee 34). But the decision of most mainstream newspapers to put a picture taken on a mobile phone of a dead Gaddafi on their front pages led to considerable debate among journalists.

And yet the work of Zelizer (2010) and Hanusch (2012, 2013) reveals such images are in fact extremely rare in mainstream media. Both find that the likelihood of such 'death' photos tend to be implied/presumed, fractional or masked, rather than overt. And in Israeli newspapers there has been a step back from graphic representations of death since the bus line Number 5 attack in 1994 (Morse, 2014).

In his study of coverage of the Haitian earthquake, Hanusch (2012) found Anglo-American newspapers pale in comparison to the kind of material to which Latin American audiences are exposed. Fahmy (2013) has also noted the common use of heart-wrenching videos and stills in Arabic media—and the justification that this is telling the story of the human cost, most obviously in the context of war.

Despite the increasing availability of graphic images on the internet, editors and reporters at mainstream media outlets were still very reluctant to use them. The 'breakfast table test' was still applied regardless of what was out there in cyberspace. The British 'watershed'[14] was used to make a judgement. A foreign affairs editor at a commercial organisation added that this was truer of British organisations than others:

> The main five British TV news organisations, I think, are pretty much all in the same place. We show less than the Spanish, we show less than the French, we certainly show less than the Arab and the Asian networks—there, there's no holds barred, charred bodies and the lot. We edit round it and where we do sometimes push, we put health warnings. I really don't think that we show more than say ten years ago.
>
> (Interviewee 40)

Zelizer (2010) talks about how "about-to-die" images are used in news media, and how this has been challenged by citizen witnesses in cases like the death of Neda Agha-Soltan in June 2009, during election demonstrations in Tehran. While mainstream media used the footage of Agha-Soltan's death, they blurred details, froze at salient moments or stopped the footage before the moment of death. The videos on YouTube, in contrast, showed all the details of her demise. For many editors, though, there was nothing out of place about this:

> With UGC . . . you show bodies where you see the outline of a dead body but you don't see the face, you don't show the moment of death. And as for injuries, you don't really want to see something that's so graphic that it would upset a child who might be watching at the time. But at the same time, we don't expect children to be watching without their parents and we have to openly say "what you are about to see is distressing" or try and put in as many warnings as possible in the intro but also in the piece.
>
> (Interviewee 35, editor flagship news programme, public sector broadcaster)

One of the Sunday broadsheet foreign editors said that he applied the same tests of taste to a YouTube video put on the newspaper's website as to the paper itself. On one occasion they had used footage from a Syrian town where rebels had thrown a man accused of being a traitor out of a window of a very tall building; the paper used the footage where he was begging for his life but cut it off at the point where he was dangling from the window, rather than falling to the ground (Interviewee 20). Unsurprisingly, there was a real division between editors (who make the decisions about what will eventually go into a newspaper/on screen/online) and reporters out in the field. Editors were traditionally much more cautious, while reporters who did not have the final responsibility for content were far more gung-ho. The international editor for a commercial broadcaster pointed out that several famous UGC examples did in fact show the moment of death.

> [The tape of] President Kennedy is user-generated content, as was the famous video of Concorde on fire and the multiple shots of the World Trade Centre. People are dying at that very moment and we used it because it is an important event. . . . But yes, you would have these ethical discussions, whether it's right to use certain things. I would hesitate to use some amazing footage of a flood going through a village after a natural disaster, and you can clearly see people in the water who are obviously drowning. I would have questions about using that. There would have to be a really good reason to use it, it's not just "hey, look at this, these people are drowning, wow!" That would be disgusting, frankly.
>
> (Interviewee 38)

Several reporters thought that there had actually been a decrease in the use of violent and graphic imagery in the news, which was inhibiting the way that news stories were framed and explained. The previous interviewee recalled growing up in Northern Ireland and seeing a news story after a series of bombings in Belfast (known as Bloody Friday) in which people were shovelling body parts into black plastic bags.

> That was an extremely brave decision made by some extremely good broadcast journalist at the time, but it showed what had happened, it showed what terrorism is, and I do think at times we have to show war is not a Hollywood thing where you fire a gun and someone falls over bloodlessly. We need to be able to show people reality. . . . People have to see what their taxes pay for, what their army does, what's done to their army, what bombs from 10,000 feet do to people, and if necessary at times what natural disasters are, what the world is made of. That's our responsibility, and I think that we've become way too squeamish.
>
> (Interviewee 38)

Interviewee 27, a Sunday broadsheet foreign editor, thought there was 'routine' sensitivity, however, particularly if anyone involved in a disaster was

British. This was a point of view echoed both by Interviewee 19, who said he strongly disagreed with his paper's policy of not running pictures of dead people in the paper ("I don't think we should be explicit for the sake of it, but I don't think we should pretend that people aren't being killed"), and by a broadsheet foreign editor, who used the example of Gaddafi's death recorded on mobile phones:

> I think it was important to use it because the manner of his treatment by that group of people was a really critical part of understanding their view of his reign and you couldn't understand that story if you didn't see it. The story was about the existence of the video, so it was weird not to do it. I thought we got it right and others didn't; we had a series of images quite small and I think size makes a weirdly big difference to how that feels as a presentation. It felt like you were telling a story rather than just ogling someone's death.
>
> (Interviewee 24)

Aid Agencies and UGC

If, as the previous sections showed, the attitudes of journalists to user-generated content had been cautious acceptance, the attitude of aid agencies that were interviewed was generally fear and confusion. But they did accept that UGC and social media were now integral parts of the humanitarian landscape, most notably because of the 2010 Haitian earthquake.

The quake was dubbed the first 'Twitter disaster' (Keating, quoted in Bruno, 2011:66) because of the significant role social media played. According to one Twitter-tracking service, 2.3 million tweets included the words 'Haiti' or 'Red Cross' between 12 and 14 January (Evans, 2010). And in the absence of reliable landline communications, both fundraising and emergency work were organised via social media.

According to the New Media Index, produced by the Pew Research Centre's Project for Excellence in Journalism (PEJ), from 11 to 15 January 43 percent of the news links in blogs were about Haiti, with that table reaching 82 percent on the Thursday and Friday after the quake (Pew Research Center, 2010).

But, as the PEJ points out, two different levels of activity emerged on Twitter. In terms of posts that linked to other reports, Haiti was the second largest topic, receiving 15 percent of the news links for the entire week (and 36 percent for the Wednesday and Thursday). But another phenomenon was occurring as well—tweets that didn't link people to fuller news accounts, but instead gave them direct access to take action, usually by donating money or supplies.

This method of asking for online donations was seen in further research by Nielsen and Facebook towards the end of the month. It found that 39 percent of Facebook users in the US, UK and Australia said they had donated to the Haiti appeal, 13 percent via SMS (Nielsen, 2010).

The night of the earthquake, Patrick Meier from the Harvard Humanitarian Initiative called the leading technology developer at Ushahidi—a crowdsourcing mash-up[15] set up to map post-election violence in Kenya in 2007—and they had a website covering the quake up and running within an hour. The volume of information flooding into the site—first about damage and missing persons, then about hunger, water shortages and health—proved so overwhelming that Meier called for help from Tufts University, and 300 students volunteered (Cooper, 2011).

Aid agencies who had not previously operated a 24-hour media presence said the success of the site had made them rethink their attitude to UGC and social media. The ability of these new technologies to combine the dual aims of most aid agencies—fundraising and awareness raising—meant that there was increasing pressure to take notice of what was going on in the online world.

This also came with some unexpected responsibility, however. Aid agencies' media offices saw Twitter primarily as a way of getting in touch with journalists and raising money, but one media manager at a major agency said people claiming to need help or aid also started using it to contact her agency.

> The most disturbing [tweet] we came across was somebody who said they were still under the rubble. So we got hold of the people from search-and-rescue organisations we knew were there, and let them have the details of where that person was. But it's unnerving, you have this responsibility and it could be something serious, you never know what you're going to do.
>
> (Interviewee 5)

The relationship between coverage and money raised was also being complicated by social media, with old-style evaluations of how much income could be generated being thrown out:

> I do think that the whole Twitter generation has kind of thrown out our old-fashioned understanding of evaluation and reach, because it only takes one Stephen Fry to retweet what [our agency] has said, and boom, you've reached five million people, so that whole analysis has to be rethought. . . . There's also the question of what the hit rate is. You get so many tweets, what does it matter? The rate of how many people will click on a link from Twitter to give money is obviously really low, I suppose, so it's kind of part of the whole awareness-raising, fundraising efforts, I suppose.
>
> (Interviewee 6, head of world news unit, aid agency)

Many realised the old-style press office had to change to reflect what needed to be a 24-hour operation, no matter how small the agency was. Often the burden fell on a few members of staff but, after Haiti, there was an acknowledgement that there had to be comprehensive out-of-hours cover. One head of media and corporate communications spoke of her horror when she realised that no one in her digital team had out-of-hours working in their job description, so she had to ask them for updates during the emergency as a 'favour' (Interviewee 17).

All the aid agencies had established protocols to follow during disasters. In some cases these dictated what would happen from hour to hour. For example, one agency's protocol said that someone on the ground had to tweet within four hours of a disaster striking to establish a line of communication with the public and journalists as quickly as possible. But the protocols often couldn't be followed because of the disruption caused by a sudden disaster. The Haiti earthquake, for example, brought down many channels of communication. In fact, using conventional methods of content gathering, such as employing a professional photographer, were often more about guaranteeing the pictures could be sent than the need to take them in the first place.

Still, for many, Haiti was certainly a moment when aid agencies had to rethink their attitude to social media. For others, Haiti raised questions about the efficacy of traditional approaches versus the value of user-generated content and social media.

> These days we have to do both [mainstream and social media], and the reason for that is that you don't know which one works best. One particular journalist I know at World Vision, for instance, got a story placed through direct tweeting of a particular journalist. That might work once in 100 times or it might work 80 out of 100 times, I just don't know, so we'll do both.
>
> (Interviewee 8, head of media, aid agency)

While one media manager took a picture of a whole load of boxes of aid covered in tarpaulins, which became one of her agency's most retweeted images; she felt, however, that there was possibly too much expected from user-generated content, given the difficulties of gathering it (Interviewee 5).

There was also general confusion about how to raise awareness via social media. Aid agencies had not fully got to grips with who their audience was, sending out messages that would be seen both by the media and the general public, all within the scope of limited characters. Many aid agencies tried to get around this by having several different Twitter accounts targeted at different audiences. So, for example, Tearfund had its official Twitter account and its 'Transform' Twitter account aimed at young people, and every member of the press office had a personal account identifying themselves as a Tearfund employee.

For the head of media relations at a major aid agency, it was clear that different social networks fulfilled different roles and, as such, there was clear demarcation of what should be used in which network:

> I think Twitter is a professionals' network, I think Facebook is a supporters' network. . . . It's quite interesting, my followers on Twitter, they're almost as if they're in distinct sub-categories, so there's a whole load of NGOs, international and national and small NGOs. There's some journalists, but I notice journalists speak to other journalists, they don't tend to go outside their journalism circle. There's a few politicians, who I think are child sponsors, mainly, and then there's some professional anti-aid people, who I get into arguments with occasionally.
>
> (Interviewee 1)

There was pressure on aid agencies to be active on social media; they were not always clear how effective this was. "[We are] tweeting, yeah, all the bloody time. Constantly . . . it gets a bit obsessive at times", said the senior humanitarian press officer at one agency (Interviewee 13).

For mid-sized agencies, there was a clear need to be innovative to try to attract attention, particularly during big disasters when Disasters Emergency Committee appeals limited individual action.[16] The key to attracting journalists was offering eyewitness accounts, whatever the format.

> So for Pakistan, our country rep lives in Pakistan and is Pakistani, and his mother is British, so his English is perfect and he understands UK media. So we wrote a blog with him to go to Channel 4. We had somebody working in the press office who'd done radio, and she recorded an interview and put it on the website. Suddenly we had collateral that we could offer out and show people.
>
> (Interviewee 17, head of media and corporate communications)

In many cases, the need for the personal was so strong that aid agency field staff would tweet emotional stories that they themselves had experienced.

> During the bombardment of Gaza at the end of November [2012], we relied a lot on a guy who is the press officer for [our agency] in Palestine. He lives in Gaza City so he was tweeting daily about "all my windows were blown out today, there were shards of glass all over my little girl because of a bomb next door". We had that unique combination of someone on the ground involved in the aid effort who was also personally affected by the emergency, and when you have someone like that, I think that's always best from a media perspective.
>
> (Interviewee 10, media and advocacy manager)

There was some confusion, however, as to which of the social media networks aid agencies should focus on; one agency employed a specific multimedia press officer to decide which technologies the news operation should use. For aid agencies with a younger supporter base, though, social media had become a necessity. For example, Islamic Relief had a relatively young base and saw both its Twitter and Facebook followers double within 18 months.[17]

While many aid agencies talked about engaging with the public, what they found particularly useful was the way social media let them target journalists. For example, Interviewee 13 recalled seeing the *Channel 4 News* correspondent Jonathan Miller tweeting that he was in Jordan. He said he immediately contacted the person responsible for the agency's media communications in Jordan, to ask her to get in touch with Miller.

For the smaller agencies, the supposed egalitarianism of social media was not working; they felt even more squeezed out in the media noise. A digital manager for one of the smaller DEC agencies pointed out it was limited to Facebook and Twitter, and had no capacity to do photo galleries or video on its website.

> I mean, it's hard without brand recognition because you don't have that trust, whereas if you're looking at a Twitter feed and you see Oxfam, you've got that immediate recognition and you're going to take the time to read their tweet, whereas we're kind of fighting that battle at the moment.
>
> (Interviewee 4)

Aid Agencies and Ethics of Social Media Use

Most aid agencies said they were uneasily aware that user-generated content and social media use threw up ethical questions, and admitted they were still in the process of working out protocols for dealing with them. Many said they were relying on codes of conduct drawn up with traditional media in mind. One drew up social media guidelines for the use of photography when tweeting based on the International Federation of the Red Cross and Red Crescent Code of Conduct,[18] but another admitted that its sign-off procedure was the same as for mainstream media, despite the very different time lines. One said that he was concerned that staff would not see tweets in the same way as other media.

> If you take a photo and then tweet it, that to me is a photograph, therefore you should have followed the code [of conduct]. Just because you did it with your mobile phone doesn't make it any different. I think that's the key issue: how much your disclaimer "oh, this is my Twitter feed, therefore it's not [the agency]", how much of that is a get-out clause? I think that's debatable.
>
> (Interviewee 13)

Perhaps because it was seen as obvious—or perhaps the reverse—very few of the agencies mentioned consent in relation to social media, but those who did saw it as at the heart of what any ethical guidelines should be about.

Security was another big issue, with many agencies realising social media—and in particular public-facing sites like Twitter—could potentially prove a risk. One agency had covered Twitter in its security handbook, while another had written a 'toolkit' reminding employees that even on their personal Facebook or Twitter accounts, staff confidentiality had to be respected. Problems had arisen less through emergencies and more through day-to-day issues—for example, during a redundancy process, despite being warned, one member of staff posted details on their Facebook page.

As a result, one agency had made particularly clear to its staff not to breach security, confidentiality or be disrespectful; the biggest challenge as a faith-based non-proselytising agency, however, was to ensure none of its staff were seen to be proselytising on their Facebook or Twitter pages.

Graphic Images

The main problem for aid agencies was not that staff were posting inappropriate material, rather that they were reluctant to post any material at all. There were concerns, though, that guidelines could be breached in the casual, rapid world of social media. One of the major agencies appeared to have thought more than most about how to apply or modify its usual guidelines to a world of social media and user-generated content. It said it would not use dead bodies and, in a situation like the war in Syria, staff would be constantly reminded that they couldn't show people's faces and names should be changed for security reasons.

> We haven't shown anyone's faces in our pictures in Syria. If there were more staff tweeting from the field, and tweeting pictures of people with their faces, I don't know what we would do. We would have to do something about that, we would have to be very careful about that. That's a thing that should be dealt with by the guidelines that every staff member should see before they're tweeting, but obviously in an emergency situation people don't have time to be watching.
>
> (Interviewee 2, head of news, aid agency)

Several agencies were devoting considerable amounts of time and money to training up staff in social media—not just press officers, but humanitarian staff (who are the first to be sent out to disaster zones) and local aid workers on the ground.

Verification

There was tension, however, between the immediacy of networks such as Twitter and the agency's traditional and often sclerotic processes of sign-off, which proved problematic when trying to react quickly.

> We've been trying to get our humanitarian staff to tweet from the ground so we get an instant voice from the ground. We are doing a social media review of our guidelines at the moment, just to make it crystal clear what people should and shouldn't be saying . . . I think it's probably security we're most worried about, anything that could compromise safety and the ability of the programme to function. So say a situation in Goma where you have a local staff member tweeting that M23[19] are in the town, they're massacring people, and they're being funded by Rwanda, that kind of thing. That might be their impression of what's going on, but it could cause huge problems for them and the programme. They might just have not have realised that by tweeting it they're saying it to the entire world, not just to [our] supporters.
>
> (Interviewee 7, world news officer)

Scepticism About UGC

After the Haitian earthquake, despite all the hype about UGC, a major agency concentrated on taking a mainstream media journalist out as soon as possible on commission from the *Telegraph*'s Saturday magazine. But because the trip took a couple of weeks to organise, the worst of emergency had happened by the time the journalist arrived. "They were walking into the middle of not quite a shitstorm at that stage", said Interviewee 5, "a slightly more organised shitstorm, but it was messy, and very complicated."

The agency was typical of many of the bigger, more established agencies, who could rely on media attention and were much more sceptical about the use of social media and user-generated content. For them, the focus was still very much on building relationships with mainstream media in order to publicise a particular disaster. As another senior humanitarian press officer put it, while they wanted to use the mainstream media's fascination with social media, hitting the prime-time TV news slots remained the most important thing:

> If you get on the Ten O'Clock News, you get everything else as well. So we want to be on their website, we want to be on their tweets, we want to be on their correspondents' tweets, we want all that. It's best to get it all at once. If we got a *Huffington Post* piece, plus a nice audio slideshow on the BBC website, we wouldn't turn that down, but it isn't going to get the reach that we require in terms of audience, but also it won't be agenda-setting. Why was the East Africa crisis [in 2011] so well-funded, the biggest emergency response we've had from the public in Africa in a time of austerity? It's because the BBC sent a journalist out, and not just any journalist but a star reporter, to the Dadaab camp for a week.
>
> (Interviewee 13)

One of the smallest agencies was the most critical of the claims that many aid agencies were making for social media, and in particular the role it could play in raising awareness and funds during crises. As its digital media manager put it:

> Despite whatever agencies will tell you, whatever they've done on social media won't have made money, it won't have appealed to people outside their sphere. People in aid agencies are very blinkered to whom they need to talk to. They deem success as, for example: 'during a food crisis we had a lot of conversations going on social media'. That's one thing, but you want to look at who those conversations are with. 99 per cent of those conversations are Save the Children talking to Oxfam talking to Merlin, and it's basically you're preaching to the converted. Is that useful? I'd brutally say it's a waste of fucking time. . . . For us, we want to be talking to people that have a potential to share it and influence other people and ultimately reach people that will give us money for the appeal.
>
> (Interviewee 12)

Faith-based organisations also had the issue that much of their support came through churches, with information being spread during Mass, for example; as such, their supporters were more likely to be older and less likely to be engaging in social media.

For the senior humanitarian officer there was an issue around authenticity. He made the point that even if the agency website or tweets were getting as big an audience as, for example, the BBC's *Ten O'Clock News*, he still would not see the two as equally important. For him the key word was 'validation'; if the story he wanted to tell was taken up by another media outlet, it wouldn't be merely his agency telling it and the message would be taken more seriously by the public and decision makers.

> So, let's say our website got as much as the DEC appeal and was beamed into everybody's home at ten o'clock at night, I think the reaction of the audience would be "yeah, well, [that agency] would say that, wouldn't they?" . . . What using other people's platforms, or using media especially, has is this issue of validation, which is essentially saying you don't have to agree with what [the agency's] saying, in fact we're not asking you to agree with what [the agency's] saying, but what it is saying is important, and it's important because we're including it in our report, so we've validated [the agency's] message.
>
> (Interviewee 13)

He felt that there were severe limitations in aid agencies' use of social media and user-generated content, because it was outside their area of expertise. Those who followed an NGO on social media were more likely to be that organisation's supporters. But he contended that communicating with that group would only take an agency so far, and they were often far from having the skills to make social media a success:

> We are not broadcasters. We are not, even when we do it, very good at it. There will be times, like Invisible Children [the KONY 2012 campaign], there will be times when something will just take off and I don't think you can replicate that. Everybody wants it. "Ooh, it'll go viral". No, it won't bloody go viral. A sneezing panda goes viral.
>
> (Interviewee 13)

Conclusion

The journalists who were interviewed frequently expressed tensions and inconsistencies about the use of user-generated content. While editors in the main saw it as an inevitable part of journalism reporters frequently regarded it as a threat to the journalistic norms they defined themselves by, even if—in an era of financial cutbacks and technological advances—the authority of journalistic presence was already being challenged.

Despite the ubiquity of UGC on Twitter and other social networking sites, there was confusion over basic privacy or permissions practices. There was still what Wardle, Dubberley and Brown (2014) called a 'Wild West' approach and, in the crucible of the 24-hour news cycle, a hope that if permission could not be sought before publication, it could be sought afterwards.

Surprisingly, however, there were strong prevailing views that, despite more graphic images being filmed and easily transmittable (and indeed viewed on YouTube), there was a pressure to resist such advances; and that this was a way of reasserting journalistic integrity.[20]

There was no sense, however, that the journalists would classify UGC as journalism in its own right; its creators were seen as sources whose treatment was often perfunctory, unless they had sought-after footage.

Like the journalists, the aid agencies questioned saw user-generated content and the social media on which it was transmitted as being a vital part of their press operations in a post-Haiti environment. But trying to adapt to this new technology had caused them considerable confusion in applying their codes of conduct, and concern for beneficiaries in a rapid media environment. There was a feeling of having to be involved, often without a clear vision of what it was that they were trying to achieve, or indeed what was being achieved.

They did not feel they could opt out of using social media, despite the view of many of the more senior press officers that they would be better concentrating on conventional media outlets. Press office concerns about verification, taste and privacy seemed academic because they rarely found themselves with material whether from their own staff or beneficiaries. They attempted to rely on current codes of conduct rather than apply new social media guidelines.

This was, however, a rapidly changing field, and there was a determination among many agencies to look for innovative ways of interacting with supporters and beneficiaries—even if at times they were concerned about the intrinsic lack of control.

Both groups were primarily concerned with how these potential new players in the field affected them, and the importance of demarcating the lines of journalistic authority. Neither group talked, unless specifically asked, about the UGC creators themselves; even then, they talked in terms of copyright and consent (the journalists), and security (but often security of the aid agency rather than that of the survivors).

The increasing ease with which photographs, copy and videos can be published and sent via social media, however, means that both the media and aid agencies are exploring areas outside this traditional dichotomous relationship. These are the subjects of the next two chapters.

Aid agencies, as significant players in the field, are challenged by UGC—it allows survivors and beneficiaries to reach an audience not circumscribed by the agencies themselves. Chapter 5, therefore, deals with how aid agencies have repositioned themselves within the field. It looks at how they have started to co-opt and control user-generated content, in order to retain their own authority in the reporting of humanitarian crises. In particular it looks at the use of Twitter, bloggers and social media influencers and the setting up of Save the Children UK's global media unit to control a humanitarian story across country borders, so theirs is the voice and the content that the media turns to in times of crisis, with ready-packaged material and a local angle.

Notes

1. The debate has long raged about whether journalism is a trade or a profession—does a person need a qualification to call themself a journalist? See Tumber and Prentoulis (2005) for example.
2. A new convention of initial reporting of breaking news stories has become the news story, which basically curates tweets—for example the Haiti earthquake (Simon, 2010) and the death of Mandela (Gayle, 2013).
3. Interviewee 21 used the example that an Iranian earthquake measuring 7.6 on the Richter scale in Saravan, April 2013, was the strongest in the country for the past 40 years, but because of lack of access, undeveloped media structures and limited internet, there was next to no coverage.
4. Interviewee 28 was talking about a virtual equivalent of the 'deathknock'—trying to get hold of survivors or the families of the bereaved.
5. In the case of Interviewees 36, 38 and 40, their 'editor' title is an honorific reflecting their seniority as broadcast journalists, but they are in fact reporters.
6. A 'collect' refers to a photograph (literally 'a photograph to be collected'). Journalists covering a news story would ask friends or relatives for pictures of the deceased/affected person in order to illustrate their story—or in some cases even extract photographs from a house without permission.
7. Such as the BBC UGC Hub or GuardianWitness.
8. For example, Interviewee 37 recalled that in Libya there had been coverage of the funeral of a baby purportedly killed in an airstrike, but someone from the

hospital contacted the team an hour before broadcast to say that it had been a miscarriage not a military attack. Meanwhile, Interviewee 20 had nearly put up video of what was said to be evidence of pro-Assad atrocities in Syria, but which had turned out to be film from the Mexican drug war; he said he had been assured by activists that it was genuine, but because he thought the heads looked 'a bit Latin American' [sic], they had gone through a more rigorous checking process.

9. The BBC had spotted the photograph on Twitter and Hamilton said that the team had had some information that pointed to its veracity, while still putting a disclaimer on the picture. It was discovered to have been taken ten years earlier in Iraq by professional photographer Marco di Lauro.
10. The *Guardian*'s communities editor later went on the liveblog to apologise and confirm it had been taken down www.theguardian.com/world/blog/2011/mar/11/japan-earthquake?commentpage=3.
11. Referring to the old cliché about not wanting to know what goes into a sausage if you eat it, he is referring to the public not wanting to know how a news story has really been constructed.
12. Most commonly it was a quick @ message on Twitter to check that it would be fine to reproduce the picture.
13. This is complicated by lack of legal precedent both on Twitter and YouTube, from where much of the content is downloaded. See Chapter 8.
14. According to Ofcom, the watershed (during which content unsuitable for children can be broadcast) on free-to-air television is 9pm to 5.30am. See http://ask.ofcom.org.uk/help/television/watershed.
15. A mash-up is a web page or application that uses and combines data, presentation or functionality from two or more sources to create new services. The Ushahidi mash-up in Kenya used Google Earth technology to map incidents of crime and violence with ordinary people reporting incidents via SMS, phone and email.
16. During a so-called 'period of joint action' (usually 14 days at the peak of the appeal) all appeals for money by the DEC agencies are branded as DEC, and agencies cannot promote themselves individually.
17. Islamic Relief UK's Twitter following was at 14,000 and Facebook at 28,000 at the time of interview in 2013. By May 2018 it was at 49,000 Twitter followers and 270,000 Facebook followers.
18. See point 10 in the Code: www.ifrc.org/Global/Publications/disasters/code-of-conduct/code-english.pdf
19. The March 23 Movement, a rebel military group.
20. The change was accelerated after the Islamic State group released a series of slickly-edited videos depicting the beheading of two American journalists, James Foley and Steven Sotloff, and David Haines, a British aid worker, all wearing the bright orange jumpsuits associated with the Guantanamo Bay captives. After the announcement of Sotloff's death, Al Jazeera PR put out the following tweet:

    We respect Steven Sotloff and won't air images of his death, or him in a jumpsuit. We suggest all media do the same. #ISISmediaBlackout—Al Jazeera PR (@AlJazeera) September 2, 2014

    It was only by the time aid worker Alan Hennings was killed in October 2014 that more mainstream newspapers started to use images of Haines other than the jumpsuit one, assisted by a widely used and attractive picture of Hennings with a child, that had been handed out by his family. Smith (2014) quotes Lloyd Embley, editor-in-chief at the *Mirror* group as saying, "They [ISIS] can't win", although Smith comments that it took four killings in six weeks for editors to realise that they had come close to "colluding" with torture.

References

Boltanski, L. 2000, "The legitimacy of humanitarian actions and their media representation: The case of France", *Ethical Perspectives*, vol. 7, no. 1, pp. 3–16.

Brown, P. 2015, 25 August-last update, *OMG I can't ever unsee that: What happened when the live aftermath of the Bangkok blast was streamed on Periscope* [Homepage of Eyewitness Media Hub], [Online]. Available: https://medium.com/1st-draft/omg-i-can-t-ever-unsee-that-what-happened-when-the-aftermath-of-the-bangkok-bomb-blast-was-7a3f39ee2b0 [2016, 7 January].

Bruno, N. 2011, *Tweet first, verify later*, Reuters Institute for the Study of Journalism, Oxford.

Bruns, A. 2008, *3.1. the active audience: Transforming journalism from gatekeeping to gatewatching*. Available: http://snurb.info/files/The%20Active%20Audience.pdf [2018, 2 July].

Burgess, J., Vis, F. & Bruns, A. 2012, 6 November-last update, *Hurricane Sandy: The most Tweeted pictures* [Homepage of The Guardian], [Online]. Available: www.guardian.co.uk/news/datablog/gallery/2012/nov/06/hurricane-sandy-tweeted-pictures [2014, 14 September].

Burrell, I. 2005, 3 January-last update, *On the front line* [Homepage of The Independent], [Online]. Available: www.independent.co.uk/news/media/on-the-front-line-15131.html [2015, 8 January].

Carlson, M. 2015, "The many boundaries of journalism", in *Boundaries of journalism: Professionalism, practices and participation*, eds. M. Carlson & S.C. Lewis, Routledge, Oxford, pp. 1–19.

Carlson, M. & Lewis, S.C. 2015, *Boundaries of journalism: Professionalism, practices and participation*, Routledge, Oxford.

Channel 4 News. 2012, 27 March-last update, *Syria's video journalists battle to tell the 'truth'* [Homepage of Channel 4 News], [Online]. Available: www.channel4.com/news/syrias-video-journalists-battle-to-tell-the-truth [2015, 6 August].

Chouliaraki, L. 2006, *The spectatorship of suffering*, Sage Publications, London.

Chouliaraki, L. 2008, 27 February, *Distant suffering in the media: Inaugural lecture at the LSE*, LSE, London. Available: http://richmedia.lse.ac.uk/publicLecturesAndEvents/20080227_1830_distantSufferingInTheMedia.mp3 [2018, 2 July].

Cooper, G. 2011, *From their own correspondent? New media and the changes in disaster coverage: Lessons to be learned*, Reuters Institute for the Study of Journalism, Oxford.

Cooper, G. 2012, "Facing up to the ethical uses of Facebook", in *The phone hacking scandal: Journalism on trial*, eds. R. Keeble & J. Mair, Abramis, Bury St. Edmunds, pp. 353–230.

Deuze, M. 2007, *Media work*, Polity Press, Malden, MA.

Duncan, S. & Newton, J. 2014, "Emotion and trauma in reporting disaster and tragedy", in *The future of humanitarian reporting*, ed. G. Cooper, City University, London, pp. 69–74.

Evans, M. 2010, 15 January-last update, *Haitian earthquake dominates Twitter* [Homepage of Sysomos], [Online]. Available: http://blog.sysomos.com/2010/01/15/haitian-earthquake-dominates-twitter/ [2012, 17 May].

Fahmy, S. 2013, 17–21 June, *Mapping the visual coverage of death*, ICA, London.

Frosh, P. & Pinchevski, A. 2009, "Crisis-readiness and media witnessing", *The Communication Review*, vol. 12, no. 3, pp. 295–304.

Furness, H. 2012, 27 May-last update, *BBC news uses 'Iraq photo to illustrate Syrian massacre'* [Homepage of The Telegraph], [Online]. Available: www.telegraph.co.uk/culture/tvandradio/bbc/9293620/BBC-News-uses-Iraq-photo-to-illustrate-Syrian-massacre.html [2015, 5 November].

Gans, H.J. 1980, *Deciding what's news: A study of CBS evening news, NBC nightly news, 'Newsweek' and 'Time'*, Constable, London.

Gayle, D. 2013, 6 December-last update, *How the world found out Mandela was dead* [Homepage of Mail Online], [Online]. Available: www.dailymail.co.uk/news/article-2519484/How-world-Mandela-dead-Tributes-flood-Twitter-users-mark-passing-South-Africas-liberation-hero.html [2015, 17 June].

Gieryn, T.F. 1999, *Cultural boundaries of science: Credibility on the line*, University of Chicago Press, Chicago and London.

Gillmor, D. 2005, 6 January-last update, *Tsunami and citizen journalism's first draft* [Homepage of Dan Gillmor], [Online]. Available: http://dangillmor.typepad.com/dan_gillmor_on_grassroots/2005/01/tsunami_and_cit.html [2012, 16 May].

Glocer, T. 2006, 11 October-last update, *We media speech* [Homepage of Tom Glocer], [Online]. Available: http://tomglocer.com/blogs/sample_weblog/archive/2006/10/11/98.aspx [2012, 20 May].

GNM Press Office. 2009, 23 June-last update, *MPs' expenses: The Guardian launches major crowdsourcing experiment* [Homepage of The Guardian], [Online]. Available: www.theguardian.com/gnm-press-office/crowdsourcing-mps-expenses [2015, 7 August].

Halverson, N. 2012, 30 October-last update, *Top 5 fake hurricane Sandy photos* [Homepage of Discovery News], [Online]. Available: http://news.discovery.com/tech/top-5-fake-hurricane-sandy-photos-121030.htm [2015, 7 August].

Hamilton, C. 2012, 29 May-last update, *Houla Massacre picture mistake* [Homepage of BBC], [Online]. Available: www.bbc.co.uk/blogs/theeditors/2012/05/houla_massacre_picture_mistake.html [2015, 5 November].

Hanusch, F. 2013, "Sensationalizing death? Graphic disaster images in the tabloid and broadsheet press", *European Journal of Communication*, vol. 28, no. 5, pp. 497–513.

Hanusch, F. 2012, "The visibility of disaster deaths in news images: A comparison of newspapers from 15 countries", *International Communication Gazette*, vol. 74, no. 7, pp. 655–672.

Harding, P. 2010, *The great global switch-off: International coverage in UK public service broadcasting*, POLIS and Oxfam, London.

Hermida, A. & Thurman, N. 2008, "A clash of cultures", *Journalism Practice*, vol. 2, no. 3, pp. 343–356.

Hofstetter, B. & Schönhagen, P. 2014, "Changing newsroom structures and journalistic routines", *Studies in Communication/Media*, vol. 2, no. 5, pp. 228–252.

Jamieson, K.H. & Waldman, P. 2004, *The press effect: Politicians, journalists, and the stories that shape the political world*, Oxford University Press, New York.

Jarvis, J. 2009, 7 June-last update, *Product vs process journalism* [Homepage of Buzzmachine.com], [Online]. Available: http://buzzmachine.com/2009/06/07/processjournalism/ [2018, 9 May].

Jenkins, H. 2006, *Convergence culture: Where old and new media collide*, New York University Press, New York and London.

Kennicott, P. 2010, "Graphic visuals in Haiti coverage break from tradition", *The Washington Post*, Washington DC, 16 January, p. C10.

Laird, S. 2012, 29 October-last update, *Incredible viral soldier pic debunked by the military* [Homepage of Mashable], [Online]. Available: http://mashable.com/2012/10/29/viral-soldier-pic-debunked/ [2015, 7 August].

Moore, H. 2012, 31 October-last update, *Even a superstorm is no excuse for journalists not to check Twitter trolling* [Homepage of The Guardian], [Online]. Available: www.theguardian.com/commentisfree/2012/oct/31/superstorm-journalists-check-twitter-troll [2015, 7 August].

Morse, T. 2014, "Death images in Israeli newspapers: Ethics and praxis", *Journalism Studies*, vol. 15, no. 1, pp. 99–113.

Newman, N. 2011, *Mainstream media and the distribution of news in the age of social discovery*, Reuters Institute for the Study of Journalism, Oxford.

Newton, J. & Duncan, S. 2012, "Hacking into tragedy: Exploring the ethics of death reporting in the social media age", in *The phone hacking scanda: Journalism on trial*, eds. R. Keeble & J. Mair, Abramis, Bury St. Edmunds, pp. 208–219.

Ngak, C. 2012, 31 October-last update, *Twitter users @comfortablysmug apologises for Sandy hoax* [Homepage of CBS News], [Online]. Available: www.cbsnews.com/news/twitter-user-comfortablysmug-apologizes-for-sandy-hoax/ [2015, 7 August].

Nielsen Wire. 2010, 22 January-last update, *Survey: Global Facebook users generous texters for Haiti, but unprepared themselves* [Homepage of Nielsen Wire], [Online]. Available: www.nielsen.com/us/en/insights/news/2010/survey-global-facebook-users-generous-texters-for-haiti-but-unprepared-themselves.html [2015, 5 August].

Niles, R. 2005, 12 July-last update, *Did London bombings turn citizen journalists into citizen paparazzi?* [Homepage of Online Journalism Review (USC)], [Online]. Available: www.ojr.org/050712glaser [2018, 9 May].

Pew Research Center. 2010, 20 January-last update, *Social media aid the Haiti relief effort* [Homepage of Pew Research Center], [Online]. Available: www.journalism.org/2010/01/21/social-media-aid-haiti-relief-effort/ [2015, 5 August].

Sambrook, R., 2010, *Are foreign correspondents redundant? The changing face of international news*, Reuters Institute for the Study of Journalism, Oxford.

Satter, R. 2015, 11 January-last update, *Witness to Paris officer's death regrets video* [Homepage of Associated Press], [Online]. Available: http://bigstory.ap.org/article/5e1ee93021b941629186882f03f1bb79/ap-exclusive-witness-paris-officers-death-regrets-video [2015, 2 November].

Schiller, D. 1979, "An historical approach to objectivity and professionalism in American news reporting", *Journal of Communication*, vol. 29, no. 4, pp. 46–57.

Schudson, M. 2001, "The objectivity norm in American journalism*", *Journalism*, vol. 2, no. 2, pp. 149–170.

Schudson, M. & Anderson, C. 2009, "Objectivity, professionalism, and truth seeking in journalism", *The handbook of journalism studies*, eds. K. Wahl-Jorgensen & T. Hanitzch, Routledge, New York. pp. 88–101.

Seaton, J. 2005, *Carnage and the media: The making and breaking of news about violence*, 1st edn, Allen Lane, London.

Shapiro, I., Brin, C., Bedard-Brule, I. & Mychajlowycz, K. 2013, "Verification as a strategical ritual: How journalists retrospectively describe processes for ensuring accuracy", *Journalism Practice*, vol. 7, no. 6, pp. 657–673.

Silverman, C. 2012, 28 May-last update, *BBC mistakenly uses image of Iraq in Syrian massacre story* [Homepage of Poynter], [Online]. Available: www.poynter.org/news/bbc-mistakenly-uses-image-iraq-syrian-massacre-story [2015, 5 November].

Silverman, C. (ed). 2014, *Verification handbook: A definitive guide to verifying digital content for emergency coverage*. European Journalism Centre. Available: https://verificationhandbook.com/downloads/verification.handbook.pdf [2018, 2 July].

Simon, M. 2010, 14 January-last update, *Haiti: what we're hearing via social media* [Homepage of CNN], [Online]. Available: http://edition.cnn.com/2010/TECH/01/13/haiti.social.media/ [2015, 17 June].

Singer, J.B. 2014, "User-generated visibility: Secondary gatekeeping in a shared media space", *New Media and Society*, vol. 16, no. 1, pp. 55–73.

Smith, J. 2014, 5 October-last update, [Homepage of The Independent], [Online]. Available: www.independent.co.uk/voices/comment/alan-henning-beheading-the-media-finally-stops-playing-the-terrorists-tune-9774984.html [2015, 10 August].

Snopes. 2017, 21 September-last update, *Did the red cross steal donations in Texas, empowered by an Obama-Era executive order?* [Homepage of Snopes], [Online]. Available: www.snopes.com/fact-check/red-cross-steals-texas/ [2018, 9 May].

Tenenboim-Weinblatt, K. 2009, "Rethinking truth through truthiness", in *The changing faces of journalism: Tabloidization, technology and truthiness*, ed. B. Zelizer, Routledge, London and New York, pp. 101–104.

Tuchman, G. 1972, *Making news: A study in the construction of reality*, Free Press, New York.

Tumber, H. & Prentoulis, M. 2005, "Journalism and the making of a profession", in *Making journalists: Diverse models, global issues*, ed. H. de Burgh, Routledge, London, pp. 58–73.

Viner, K. 2009, 15 June-last update, *Internet has changed foreign policy for ever, says Gordon Brown* [Homepage of The Guardian], [Online]. Available: www.theguardian.com/politics/2009/jun/19/gordon-brown-internet-foreign-policy [2015, 6 August].

Wahl-Jorgensen, K., Williams, A. & Wardle, C. 2010, "Audience views on user-generated content: Exploring the value of news from the bottom up", *Northern Lights: Film & Media Studies Yearbook*, vol. 8, no. 1, pp. 177–194.

Wardle, C., Dubberley, S. & Brown, P. 2014, *Amateur footage: A global study of user-generated content in TV and online news output*, Tow Center for Digital Journalism, New York.

Whittle, S. & Cooper, G. 2009, *Privacy, probity and public interest*, Reuters Institute for the Study of Journalism, Oxford.

Williams, A., Wardle, C. & Wahl-Jorgensen, K. 2011, "Have they got news for us? Audience revolution or business as usual at the BBC", *Journalism Practice*, vol. 5, no. 1, pp. 85–99.

Zelizer, B. 2010, *About to die*, 1st edn, Oxford University Press, Oxford.

5 Twitter Takeovers, Blogging and 'Following the Sun'

How Aid Agencies Use the Tropes of Social Media and User-Generated Content

> Underpinning this move to social media and the internet . . . is a sense of empowerment that goes to our donors, who can see exactly what we're doing when we're doing it, but also to the beneficiaries who are often on smartphones, dongles with internet in [refugee] camps, people are able to be part of that online conversation in lots of surprising ways and what we're saying in real time. There's a power disparity and social media is not going to solve that totally . . . but it goes to redress that balance a little bit.
>
> (Interviewee 15, acting head of news)

At the peak of the Live 8 concert in 2005, the pop star Madonna provided one of the most memorable images of the day—holding hands with a young Ethiopian woman called Birhan Woldu (Carr-Brown, Fielding and Clifford, 2005).

Woldu had been the icon of the 1984–85 famine in Ethiopia; her dangerously emaciated form had appeared in the BBC's news coverage of the crisis and most pertinently at Live Aid in 1985, as the haunting sound of The Cars' *Drive* played.

By 2005, Woldu was not only a symbol of hope but also, as the video was replayed 20 years on, had become an uncomfortable reminder of the kind of reporting and imagery that had been used widely at the time, by both journalists and aid agencies, to shame the developed world into helping people affected by the famine (Bristow, 2005). The disquiet such images caused among NGOs saw Oxfam produce the *Images of Africa* report (van der Gaag and Nash, 1987). It talked of the use of 'truly pornographic' images of famine, and ultimately led to the creation of the Red Cross Code of Conduct in Disaster Relief in 1992.[1]

Today, the conclusions of the Red Cross Code—that those caught up in disasters should be portrayed as dignified human beings and not 'hopeless objects'—may seem obvious. The insensitive nature of some of the NGOs' fundraising and messaging in the 1970s and 1980s (Benthall, 1993) has been challenged and largely changed, though Dogra (2012) points out the continuing decontextualisation of poverty in NGO imagery. Others have examined ongoing issues of 'difference' and 'distance' in NGOs' depiction of 'the marginalised' in their communications materials (Orgad and Seu, 2014).

Has the increasing use of digital media technologies made any difference to the way NGOs can bring different voices to the fore in humanitarian crises? For some of the aid agencies featured in this book, it has meant trying new approaches. For example, Oxfam GB's 'Twitter takeover', which allowed a Syrian refugee to write unmediated tweets on its account; Save the Children's #hiddencrisis link up with the social media news agency Storyful; Christian Aid's link up with On Our Radar SMS project and Plan UK's trip to Ethiopia accompanied by a blogger rather than a journalist.

In such cases, the aid agencies are going beyond their traditional roles of primary source (Anderson, 1993 1991, 1997; Schlesinger, 1978; Schlesinger and Tumber, 1994) or gatekeeper (Beckett, 2009). Instead, they potentially become a type of mediator, using social media and inexpensive technology to afford a voice to the previously voiceless.

But taking on this new role is no easy task, as this chapter shows. NGOs have found themselves caught up in a rapidly changing media environment, and keeping staff and beneficiaries up to speed has not proved straightforward. And for those richest in resources like Save the Children UK, new technology has provided an opportunity to control the message more rigidly—not just at a national but international level.

Employing New Technologies

The UK aid agencies continually emphasised—in their mission statements, evaluation policies and media work—the need to include marginalised voices. There was a commitment to hear from 'real' voices and, as CARE International put it, even "contrary opinions" (2013:2). It was no longer seen as acceptable that agencies should speak for survivors and beneficiaries. NGO websites and mission statements often made this intention clear:

> [It's about] being rooted in the lives of real people, telling the true story, communicating real voices.
>
> (Christian Aid, 2018)

> The voices and views of minority, disenfranchised and other groups with perhaps contrary opinions should also be heard and considered.
>
> (CARE International, 2013:2)

This is a laudable aim, but many agencies were still grappling with how best to use user-generated content and social media to make beneficiary voices heard. For example, during the DEC's 2013 Syria appeal it used a live Twitter Q&A under the hashtag #decqs.[2] A media officer seconded to the DEC press office[3] sent out the following tweet, not entirely in jest: "If anyone who doesn't work for an aid agency could ask us a question about Syria using #decqs in the next 2 hours I will love you forever".[4]

The problem is that it tends to be the rapid onset disasters—those which commentators and NGOs have complained that journalists spend too much time on already—that tend to get most social media pick up (Moeller, 2006; Franks, 2006; CARMA, 2006). Twitter, Facebook, Instagram, Periscope and Snapchat lend themselves to the dramatic over the chronic; the 'success stories' of social media tend to be crises like Typhoon Haiyan and the Haitian earthquake, rather than the East Africa famine.

For many aid agencies, the solution was equipping press officers with the technology they needed to act like 'fireman reporters'—the approach pioneered by Christian Aid and Oxfam in the early 2000s (Cooper, 2007). While in the mid-2000s, a press officer would commonly take a laptop, a satellite phone and a digital camera to cover a disaster, as the years progressed press officers were also equipped with a BlackBerry, BGAN and Quicklink to stream live broadcasts.[5]

Even a middle-ranking agency would take a BGAN to an emergency, although Interviewee 6, head of the agency's World News Unit, said that the \$20–\$30 per minute cost meant it wasn't used unless absolutely necessary. At such medium-size agencies, press officers relied more on BlackBerries (which were often used for interviews), MacBooks and cameras. Smaller agencies relied solely on voice recorders and a photographer, or a local phone, laptop, camera and video camera operated by their own press officer.

> I think as long as you've got the smartphones, you can do it all. You might want to do Audioboo[m], audio clips, all that kind of stuff from your phone.
>
> (Interviewee 2, Head of News)

For the smallest agencies, the technology was pretty limited. Interviewee 3 said the communications officers had digital SLR cameras, but didn't know how to use them, and dongles for remote internet access—in contrast to her time at one of the major agencies when press officers were given "the whole shebang".

This approach frequently meant the agency had to pay for the outside broadcast themselves with nothing more than an aid agency credit to show for it. Not surprisingly, the more forward-looking agencies took a different tack. For example, as mentioned in Chapter 3, Save the Children UK's 2012 campaign #hiddencrisis aimed to raise awareness of the West African hunger crisis, which had not had much traction in the media. The campaign was specifically planned as a Twitter event; the agency took Neal Mann, the then Sky News digital media editor who had a huge Twitter following, out to Burkina Faso. It linked up with Storyful, to plot Mann and Save the Children's journey across the country, a good example of actively using new media tools to try to alter the conversation and interact with their audience.

Twitter Takeovers: Social Media as a Platform for Different Voices

Some NGOs have tried to provide a platform for 'different' voices. For example, on 1 February 2013, the Oxfam GB Twitter feed was handed over to Hasan—a Syrian refugee at the Zaatari camp in Jordan—for the day.[6] Hasan's tweets, which included pictures of his newborn baby Leen, were retweeted by celebrities such as Stephen Fry and Damon Albarn. Hasan talked about Leen and the joy of swapping a tent for a caravan—the advantage from his perspective being that when you live in a caravan you can lock it with a key and stand upright inside it. Oxfam counted it as a big success; not only did it repeat the exercise a month later with members of Hasan's family, but the *Guardian* then asked Hasan to repeat the exercise several months later for their Syria liveblog day.

> What made it work really well, was it was clearly unmediated, his English was good, but you know, there was lots of mistakes, but that added to the sense of realness of it.
>
> (Interviewee 13, senior humanitarian officer)

Another development was when beneficiaries themselves were able to be the reporters on a more long-term basis. One of the most successful of these was On Our Radar, a social action group set up in 2012 to help ordinary citizens become reporters during the Sierra Leone elections and which has worked with different NGOs since then, including Christian Aid during the 2016 Ghana elections.

On Our Radar was born out of a response to both Ushahidi, the non-profit tech company which developed open-source software for information collection and interactive mapping, and the infoasaid project run by Internews and BBC Media Action (Klein, 2014). While Ushahidi was effective in visualising citizens' reactions to crises, it neither constituted nor facilitated conversation or analysis beyond the crisis. In comparison the infoasaid project, with its mantra of 'communication as aid' (infoasaid.org, 2012) did facilitate conversation as did a number of other projects which came under the Communicating with Disaster Affected Communities (CDAC) Network, which was set up in 2009 and brought together more than 30 humanitarian, media development, social innovation, technology and telecommunication organisations to "make aid more effective through communication, information exchange and community engagement" (CDAC, 2018).

What was unusual about On Our Radar however was that it saw the need for reporting as well as communication—"more comprehensive communication to flow out from, as well as back into, poor or disaster-affected communities" (Klein, 2014:31) in a journalistic environment. While many commisaid projects were about information flow, On Our Radar sought to give voice to those affected by crises by turning them into citizen reporters, training them to use SMS messages as a broadcasting tool.

To report on the Sierra Leonean elections, Libby Powell, the founder of On Our Radar, bought three mobile phones, connected to three different telecoms networks (in case of network failure), downloaded an app that could send these messages straight to an email address and then be shared via social media or direct to agencies (Halstead and Powell, 2017).

The 'reporters' trained by Powell helped highlight problems in the elections such as the failure to deploy tactile voting systems, which was then picked up by the *Guardian* (Banugra and Powell, 2012). But their real success came with the Ebola outbreak of 2014. The scale of the outbreak meant that many parts of the country were inaccessible for international journalists and public meetings were banned: the result was that credibility of information was challenged.

To counter this, the citizen reporters, many of whom now had smartphones, were able to send not only text information but also audio recordings and photos via WhatsApp (Halstead and Powell, 2017). Some areas in Sierra Leone lacked sufficient electricity supply and internet connection to constantly report on the Ebola crisis; to combat this, On Our Radar devised a simple system that allowed the organisation to send bulk SMS messages about various news issues to those reporters in such villages and they could respond and add detail via text, which would later be used to construct news reports (Ma, 2017). Media organisations that went on to use the Ebola reports included *The Guardian*, the *Huffington Post*, BBC World Service, Sky News and Channel 4 (Guardian Readers and Bannock, 2014; Kortu, 2014; Bayoh, 2014) as well as a collaberation between On Our Radar and the *New Internationalist* after the crisis was over, to examine the aftermath (Myles, 2016).

With these ventures, part of the attraction for readers was the use of ordinary, everyday pictures— for example, Hasan's children sitting in a tent and, in particular, of his baby who was born the same day as the Oxfam Twitter takeover. Typically these would be 'private' pictures taken by private individuals, but the division between private and public is being increasingly diluted (Becker, 2011; Wardle and Williams, 2008). As discussed in more detail in Chapter 8 there also is a potential breakdown in contextual integrities around privacy (Nissenbaum, 2004:119).

There were also security-related concerns about whether beneficiaries fully understand the implications of sharing information in this way. As the head of media at a large agency put it:

> Informed consent in this day and age includes making sure that the person giving consent is aware that their identity could well be seen worldwide, and that their government might well become very quickly aware of what they've said. There's no getting away from the fact that in the internet age things like that are a lot more important than they were 30 or 40 years ago, where the chances of a story even reaching a remote outpost in a war-torn hellhole were minute. These days the chances are very large that they will.
>
> (Interviewee 8)

While many agencies were positive about the Hasan Twitter takeover in general—or said that encouraging beneficiaries to record their experiences either with phones or cameras distributed by NGOs was to be commended—in reality, some were reluctant to encourage this. In the main, they explained their reluctance, not by admitting to concerns about how voices from the field could be controlled, but by saying such endeavours were impractical and ineffective compared to aid agencies' traditional method of interviewing beneficiaries themselves and then processing the quotes for journalists—something that was easier with the On Our Radar approach.

> I think anyone I met [in a disaster zone like Haiti] was at such a stage of life that if I'd asked them to take some footage for me, they probably would have laughed and said "What the ****? I don't have a house or anything". But the best thing that we could do was get them to tell me their stories, take their pictures, take video of them as they were living in the camp.
>
> (Interviewee 5, senior press officer)

> There was a project we did in Haiti, where we gave kids cameras to document their lives in the camps, and in Jordan in Zaatari camp recently, giving children and adolescents cameras to take pictures of their lives there. I think some of the material they got was really lovely, but I wonder how much investment was made, and then in terms of what was actually used at the end of the project, what the benefits were for the kids involved.
>
> (Interviewee 96, humanitarian information and communications manager)

Many aid agency press officers saw Twitter's character limit as a problem, because of the lack of space for nuance in very complex storytelling. Most harshly it was dismissed as a 'novelty package'. The acting head of news put it this way: "The story isn't what they're saying, it's *who's* saying it" (Interviewee 15). Only one agency press officer admitted that they were nervous about what might actually be said: "It's a risk—opposite to most of our work where it's about controlling the message" (Interviewee 2).

The Role of Social Media Influencers

One way NGOs have started to incorporate 'ordinary' or non-agency voices is through social media influencers. These voices are not NGO employees, are not survivors or beneficiaries, but are perceived as a step away from the mainstream media.

The impetus for this was the success of Save the Children UK's 2010 #blogladesh campaign, fronted by three 'mummy bloggers' (a popular term for mothers who blog about the minutiae of family life). Instead of journalists, the charity took Josie George (who then wrote the blog sleepisfortheweak.org.uk), Sian To (mummy-tips.com) and Eva Keoghan (nixdminx.com) to Bangladesh to raise awareness of the upcoming Millennium Development Goals conference. While out there, the women blogged, tweeted and uploaded pictures to Flickr with the #blogladesh hashtag.

The strategy appeared to reap dividends for Save the Children. The blogging site Mumsnet invited Josie George to join a web chat with the deputy prime minister Nick Clegg, and the campaign was picked up by the BBC Radio 4 *Today* programme and the ITV lunchtime news. High-profile media targets had been reached for minimal outlay; Liz Scarff, who later set up the social media consultancy Fieldcraft and the organiser of the #blogladesh campaign, estimated that the whole event had cost around £5,000—mainly on transport (Cooper, 2011, 2015).

Unsurprisingly, the success of #blogladesh led to a spate of copycat trips: Save the Children's YouTubers-driven #PassItOn in 2011; Plan UK's 2011 *Blog 4 Girls* competition; World Vision's *ShareNiger* with mummy bloggers in 2012; and Tearfund's 2013 *See for Yourself* initiative with faith bloggers.

The #blogladesh campaign had emphasised the normality of the women and their ability to relate to beneficiaries. As Scarff put it: "Who could be more powerful to tell stories about children than mothers who have their hopes and dreams for their own children?" (Cooper, 2011:32).

This was not easy to do successfully. "The concept is simple, the practice really time consuming", said Interviewee 11, a media manager at a small agency. Despite Scarff's description of it as a cheap alternative, hard-pressed agencies did not see it as an option—as Interviewee 11's colleague, the digital manager at the same agency, put it: "Take bloggers out? We can barely take ourselves out."

For those who did follow in Save the Children UK's footsteps, though, the interesting thing was that the bloggers, and then later the social media stars, were not that different from the journalists they were replacing. They were typically white, middleclass and privileged. For example, the #blogladesh mummy bloggers were all women who had previously worked in PR or had writing experience: Sian To had a specialist parenting PR company; Josie George ran a weekly writing workshop for other bloggers; and Eva Keoghan, who had previously worked for the PR consultant Lynne Franks, was now a social media consultant. #ShareNiger used To again, while #PassItOn—organised by the same consultancy—used Christine Mosler, a freelance copywriter and photographer who blogged under the name Thinly Spread; primary school teacher Lindsay Atkin, whose son was a prominent YouTuber; and political blogger Tracey Cheetham, who was also a local councillor and a director of Eclipse PR.

Faith-based agency Tearfund used bloggers with a Christian faith but otherwise they, too, were from similar white, middleclass backgrounds—one was a copywriter turned ordinand, another a cartoonist and the third a Durham academic. The only non-white blogger used by an aid agency in these early campaigns[7] was the blogger and filmmaker Waiki Harnais, who was brought up in the Democratic Republic of Congo. She won Plan UK's blogging competition in association with the *Guardian* in 2011 (Ford, 2011).

Experienced as many of these bloggers were, they had not been trained as journalists and did not, therefore, subscribe to the journalistic norms of objectivity and distance. During the Tearfund trip to Uganda, blogger Liz Clutterbuck wrote of her discomfort after a fellow blogger used a photo of her with a local child. "I felt like I was a throwback to the 19th century—a well-meaning white female missionary cuddling an African baby" (Clutterbuck, 2013). So this type of well-intentioned 'personalisation' did not always have the desired effect of breaking down the 'us and them' dichotomies.

When news stories traditionally involved 'ordinary' people in the years before social media, the result was often an opening up of access to the media to individuals, rather than social groups or organisations. Individuals would usually talk in a personal capacity about their own experience, as opposed to having a more analytical role (Manning, 2001). As mentioned in Chapter 2, Cottle (2000) argued that while 'ordinary voices' were routinely used on TV news items, they became what Beck calls "the voices of the side effects" to symbolise the human face of a news story. Whilst Cottle was talking about television interviews, his argument could equally be applied to blogging and social networking:

> Television news positions ordinary people to symbolize or (literally) stand for ordinary feelings and responses to the consequences of environmental risks not to articulate a form of 'social rationality' much less discursively challenge 'scientific rationality'. With too few exceptions the discursive play of difference and contending rational accounts is preserved for other, non-ordinary voices.
>
> (Cottle, 2000:31–32)

While #blogladesh blogger Josie George did get to meet Nick Clegg, much of the coverage focused on the experiences of the women involved. The aim was to build an emotional bridge between the mothers they met at Save the Children projects in Bangladesh and the women back in the UK who read their parenting blogs. So the Bangladesh bloggers often focused heavily on their own thoughts and feelings. Their blogs became as much about their own ability or inability to deal with the powerful impact of the sights they were exposed to as they were about the sights themselves:

> It is hard to find the words here. I didn't take pictures. Just staying upright and breathing in the space of so much . . . so much horror, and horror it was, was the best I could manage.
>
> (Josie George, sleepisfortheweak, 2 September 2010)

> What does it take to be authentic and genuine? It's not just a matter of being yourself, sometimes things find you that resonate and make you reconsider not just who you are, but what you think and feel and say, and write. And of course what you do. Bangladesh is one of those rare experiences that makes you re-examine and question humanity and as you try and go to sleep, yourself. First on the agenda is the assimilation of this place, which is hot, busy and chaotic. How do I even begin to make sense of all this?
>
> (Eva Keoghan, Nixdminx, 1 September 2010)

Such blogposts started changing the emphasis from charity beneficiary to charity blogger. As a PR (and journalistic) approach, it produced some powerful writing that undoubtedly captured the media's attention, but it risked the marginalised fading into the background again. Chouliaraki has noted that we live in a society where:

> our own private feelings are the measure against which we perceive and evaluate the world and others. [. . .] While news becomes part of this 'culture of intimacy' it implicitly allows us to focus on our own sufferings and disregard those 'others' outside our own horizon of care.
>
> (Chouliaraki, 2006:13)

So while we may look at a Facebook post and click 'Like', watch video tributes on YouTube or read tweets from refugees, this does not mean the distance between the world's poorest people and the developed world onlookers are being overcome.

Would the introduction of social media influencers—many of whom were more pictorially based than the early bloggers—make a difference? This depended on the charity involved. Islamic Relief started to use Snapchat and Instagram for important events in their calendar, such as The Cake Campaign (formerly #Cakes4Syria), a campaign started by a volunteer in Bradford in 2013 that involves buying and sharing cakes to break the Ramadan fast (Islamic Relief, 2018) and has partnered with Instagrammers to bring its message to a wider audience.

But other agencies were becoming increasingly sceptical about the use of social media influencers. One of the biggest agencies, while acknowledging the power of Instagram Stories and YouTube in particular, said they were moving back towards using 'celebrity' Instagrammers rather than 'ordinary' people. While they had used ordinary people in the past, by 2018 they had found a problem around what the agency's social media manager termed 'authenticity'.

> We're working on a complete new policy when it comes to our influencers and vloggers. There's problems—I'm sure you've seen stuff when the press go back like 10 years back on Twitter to see what people been

> tweeting I think [our agency] has our fundamental principles and we have to be really careful who we work with. Also I think it's reached saturation point—that it feels like everyone on social media has got something to flog. We don't want to look like we're just another thing they are selling.
>
> (Interviewee 98)

To combat this, the agency was working carefully with those influencers they remained connected with to ensure that any branded content 'did not look like branded content'. That again led them back to celebrities rather than 'real' people who understood the rules of the game. The agency was working with a well-established reality TV star turned TV presenter whom they trusted to share and retweet content or formulate his own rather than 'spoonfeed' him—but that was because there had been a long-term ongoing relationship with someone who was experienced in media work.

Return to Control: Save the Children's Global Media Unit

> The Global Media Unit (GMU) is a new unit within the Save the Children organisation. Based in the UK, it leads the media function across the global organisation under the networked leadership model. This is trail-blazing, demanding work, and demands a Project Officer with the requisite skills. The purpose of the Project Officer is to assist with delivering global media moments, and monitor and report on outputs from global media moments. The Project Officer will manage projects across Save the Children globally, working closing with Save the Children International (SCI) and other Save members.
>
> (Excerpt from recruitment advertisement for a project officer, August 2014, www.charityjob.co.uk/jobs/project-officer—global-media-unit/343535)

While many aid agencies were experimenting with imaginative ways to capture public and political attention through social media, Save the Children was taking on a more ambitious project—to co-ordinate media messaging on an international scale by setting up what it termed its 'Global Media Unit'.

The original Save the Children was set up in 1919 in the UK in the wake of the German and Austrian-Hungarian famine caused by the blockade by the Allies at the end of the First World War, and was instrumental in putting children's rights on the international agenda. Since then, it has grown with 30 national organisations serving around 120 countries. These national organisations come under the umbrella of the global network of Save the Children International with common bylaws and codes of conduct, although they operate independently.

Save the Children UK is currently one of the biggest and most successful NGOs in the Disasters Emergency Committee. Thirty years ago, it had a budget of £16m and was run by a retired civil servant from a school in south London (Slim, 2011). By 2016, it had an income of £405 million (a record in its 95-year history), spent £374 million on charitable activities (Save the Children annual report, 2016). Its profile had been such that in 2010 it could attract Justin Forsyth as its chief executive, a former advisor to two prime ministers Gordon Brown and Tony Blair. In recent years, Save the Children was regarded by other agencies as the most successful at getting media coverage and raising its media profile, although this enviable position was tarnished in 2018 when it became caught up in a media scandal when it emerged that both Forsyth (who had moved at Unicef in 2016) and another former senior executive Brendan Cox, widower of the MP Jo Cox, had had allegations of inappropriate behaviour made against them.[8]

Save the Children's dominance of the UK aid sector had begun however before the arrival of Forsyth in 2010, under the stewardship of Jasmine Whitbread who was CEO of Save the Children UK between 2005 and 2010, and who took over the role at Save the Children International in 2015. It already had a bigger press office than most when, in September 2012, it set up its Global Media Unit, which came under the mandate of Save the Children International but interestingly was employed, run from and line-managed by Save the Children UK, reflecting the relative power and dominance of this Save the Children member organisation.

The aim of setting up a unit which would have global reach was to ensure that messaging could be tailored around the world in the most efficient way. This was seen as an opportunity to strengthen the Save the Children brand and exploit 'media moments'[9] as outlined in the job advertisement above.

Interviewee 94, who had formerly had a senior role as a broadcast journalist and was the head of the Global Media Unit, described what he saw as the opportunity to globalise the organisation's media approach. While paying lip service to the independence of individual Save the Children members, he described the unit as a unifying force.

> What you have is an organisation which is federal . . . different countries have Save offices—[they] value their independence to do their own thing, which is fine, but in the increasingly globalised world . . . if you speak as one at the same time, you can have far more impact.
>
> (Interviewee 94, head of GMU)

So while members are known to "value their independence", the head of the GMU made clear this should be subordinate to the global reach of the Save the Children brand. For the GMU, there was less of a discussion about the conflict between fundraising and educating the public that other press offices had talked about. This was much more about reaching authority figures and corporate donations. He went on to talk in very corporate terms and jargon about the need for Save the Children to use "high impact media outreach" in order to promulgate "brand visibility". There were several reasons he mentioned this need for brand focus:

> We need visibility for the brand in order to conduct effective advocacy, much of which is behind closed doors. We also need that brand visibility in order to convene meetings, to leverage and build partnerships, and also fundraising. We need to be known and to be trusted and credible.
>
> (Interviewee 94)

The unit had been set up because there was a feeling at Save the Children that the organisation had missed a lot of opportunities for influencing decision makers such as politicians and business people and by setting up such a unit, Save the Children could position itself as the "loudest, most respected and reliable voice representing the rights and injustices of children" (Charityjob.co.uk, 2015). In particular, it was about ensuring that Save the Children were first on the ground when a humanitarian crisis occurred producing 'world-class' multimedia content, ensuring a global media strategy across different media platforms and controlling the 'Save the Children' message more completely. It was also about ensuring that messaging was not contradictory across different member organisations and the countries that they operated in.

> If we take an example like taxation, colleagues here [in the UK] will have a particularly trenchant view about taxation and how that needs to be addressed, corporates and so on, but if you talk to colleagues in our American office, they all see that differently.
>
> (Interviewee 94, GMU)

The unit looked to identify what the charity called 'shared moments' during the year that could be exploited—usually reports on subjects that were relatively uncontroversial. One key example was Save the Children's annual *State of the World's Mothers* report. The report ranked the 'best' and 'worst' places to be a mother, using data on health, education, maternal and newborn survival rates, and political instability.

For the *State of the World's Mothers 2013* report, the GMU created a text-based press release as a template, which was then adapted for different regions such as the sub-Saharan Africa, South Asia and Latin America. While this may not seem particularly innovative, Save the Children found that this copy-and-paste approach still had currency for some countries in the developing world. On top of that, there was also a multimedia component in which Save the Children staff or commissioned freelancers would go out to gather video or stills to be offered up to news organisations, which the head said were taken up in a way that would not have happened a few years earlier when he was working at the BBC.

> This has changed hugely, say, from ten years ago, 12 years ago, when, if you went to a news desk and said "we've got these materials", they would sniffily say "well, thank you very much, but actually we cherish our independence, we gather our own content". That's changed. The news model's broken, budgets are hugely under pressure, and we have an opportunity as a credible organisation—trusted organisations offer up materials that are then taken.
>
> (Interviewee 94)

The Democratic Republic of Congo (DRC) came out as the worst place to be a mother in the 2013 report, while Finland was the best. The GMU commissioned freelancers working to a set brief to film in both the DRC and Finland, along with five or six other countries, to produce 'B-roll' (privately shot footage provided free of charge) for news organisations.

> We wouldn't do a voiceover or piece to camera, but we'd try and provide all the other elements. So we would have soundbites for kids, interviewing off-camera, with subtitles or without, so they can localise it if necessary. We'd have an interview with a Save the Children staff member, we'd have cutaways, general views, we'd have all of the basic elements.
>
> (Interviewee 91, media manager, GMU)

Effectively, Save the Children's GMU was providing all the raw elements for a television or website package, leaving a mainstream media journalist with the sole job of having to voice their own script. While the DRC footage was widely disseminated, the GMU was surprised to see the Finland footage used as well.

In the UK, the footage was picked up and used by the *Telegraph's* website (which ran one piece on *Finland* and one on the *DRC*), Sky News online and the Huffington Post. The *Telegraph* put its logo on the footage, although noted that it was 'courtesy of Save the Children' right at the bottom of its web page. Sky wrapped the footage around an interview with Ishbel Matheson, then of the Save the Children press office, without indicating the footage had been taken by Save. The Huffington Post used an advertisement promo from Save on their website, without making it clear until the end it was actually a Save the Children advert. Meanwhile, the *Mirror, Daily Mail, Daily Express* and BBC ran it as a website story without a video embedded in it.

Save the Children wanted global reach, however, so the GMU came up with a list of further recommendations to individual member organisations to help them get media coverage. Interviewee 91 said that they suggested to press officers that they identify where different countries ranked on the *State of the World's Mothers* index, how each country compared with neighbouring countries and within a region, and how its place on the index had changed over time. The GMU would then write the press release with identical quotes, which could be signed off by the local chief executives.

> Traditionally what would happen before, there would be a Save US written document, and then a week or two before publication date they would just email it out and say, "this is it, do what you want". So a good press officer in any office would be able to do what we do, but it's just that they don't have the capacity, or the time. So a really important element of what the GMU does, is just to make it easy for offices, so that in some cases it's as simple as pressing forward on an email to send it to journalists.
>
> (Interviewee 91, media manager, GMU)

The media manager had gone out to Save the Children South Africa to help it organise a press event; this included inviting journalists and a local celebrity, writing briefings and Q&A sessions, and liaising with the UK celebrity team to get the James Bond actress Naomie Harris to appear in a video.

> What the main objective for us to do, apart from meeting regional correspondents, who would be useful contacts for the GMU, was to establish Save South Africa as a legitimate voice for children in South Africa, because they were basically unheard of, so it was a brand-building exercise.
>
> (Interviewee 91)

The idea of the GMU had been formulated before Interviewee 94's time at Save the Children, but he was tasked to set it up with just the help of a consultant. By July 2013 there were five members of the team—himself, two media managers, an international film and photography manager and an assistant film and photograph manager. When he arrived, he saw it as heavily influenced by the BBC's ethics and production values, because many of Save the Children's media staff had previously worked at the corporation (as had he). However he was conscious that there were many potential audiences as yet untapped and he wanted to spend more time working with Al Jazeera and Al Arabiya.

The budget for the GMU was undisclosed, but he said that: "The return on investment is significant. It's a cliché, isn't it, you have to spend to accumulate. Effectively we raise millions off the back of investing a few thousand pounds or dollars" (Interviewee 94).

In trying to describe how the GMU liaised with other parts of the organisation such as the UK media team the head of the GMU described it as a bit "Heinz 57" (using a British slang term to designate the fact that it was a mixture). He pointed out that they had got buy-in from all the Save the Children chief executives in the member organisations, but emphasised that it was a question of collaboration, not control. As Interviewee 91 put it:

> We try to set ourselves as the central spoke in that global media wheel, so we coordinate and share materials from various members with everybody else. We commission materials with an eye to how they can be used internationally.

Members of the GMU were careful to couch any description of their roles in words such as 'help' and 'assistance'; there was a consciousness that this could be seen as imposing Save the Children UK's approach on other member organisations:

> I sit here in Save UK, who are very good at media work, but I also have to be very conscious, I have to play a political game in the organisation and make sure I'm also listening intently to the other members of Save in the other countries, and take their views on board. It's not always easy.
>
> (Interviewee 94, head of the GMU)

This was reflected in how the GMU worked, and the domination of the UK office in media work. For example, *State of the World's Mothers* is a Save the Children US-led report (the US produced the text-based version), but the GMU created the multimedia kit.

Unlike many agencies, the GMU prided itself on asking media outlets to give Save the Children credit for material it had produced; the head of the GMU believed that many journalists were actually glad to give it, as it gave their report credibility and transparency. For example, one particularly successful GMU operation involved commissioning a freelance war photographer—Moises Saman—to take pictures of Syrian children asleep in refugee camps, to create a photo essay for World Refugee Day. The story was covered in outlets from *Time* to the UK's *Mail Online* to the *Sydney Morning Herald*, who all gave Save the Children a mention, although they primarily presented it as a news story about Saman's pictures.

The GMU did not just focus on pre-planned media moments, however; it was also set up to deal with perceived shortcomings in the way rapid onset emergencies had been dealt with in the past, and the need for Save the Children as a brand to be first on the ground. The GMU was tasked with establishing a way of working globally when a large-scale emergency happened. This included generating a roster of media managers from different Save the Children member organisations to be deployed in an emergency, establishing a rapid response media plan for emergencies agreed between member organisations and, most pertinently, ensuring what the GMU saw as '24-hour cover'.

> [It's] what I call the 'follow the sun' approach. It's basically passing a baton between us, then on to Save US, and on to Save Australia, so that when things blow up suddenly, we've got 24-hour cover to push out materials.
>
> (Interviewee 94)

This 24-hour approach was not just about publishing press releases or statements, but also actively ensuring that those on the ground (the humanitarian teams or local partners) would take videos and stills of the situation as soon as possible after the emergency happened. While this was still not done at the 'flick of a switch', according to Interviewee 94, he saw it as a vast improvement on the often 'sclerotic' pace that NGOs like Save the Children used to move at in the past. He explained what would happen if the equivalent of another Great East Japan quake struck. Not only would he insist on a media manager being involved in the rapid assessment team (who are flown out to emergency zones as quickly as possible), the GMU would also demand that a wider variety of material was sent back:

> We'd be thinking about all media platforms including social media. We'd be wanting to get some images back from the field ASAP, shot on iPhones. Because of social media, the concept of what we determine to be quality I think has shifted hugely in recent years and continues to shift.
>
> (Interviewee 94)

His comment about 'quality' refers to the grainy, often unprofessional looking images that UGC creators often produce. While the focus was often on employing professional freelancers in the later stages of an emergency, Fletcher realised that adopting the tropes of UGC would give a sense of urgency and authenticity, even if it had been taken by an aid worker rather than a member of the public. The focus here was on speed however—so that Save the Children could be putting up images and film as quickly as an ordinary person would.

The result was, Interviewee 94 said, that organisations who had taken the GMU's footage or done interviews with Save the Children representatives, included not only the BBC, but also wire agencies, Al Jazeera and, in particular, French and German media. In multimedia packages GMU would even offer translation of speech if necessary, although wire agencies like AFP tended to do their own.

There were clearly sensitivities about the GMU's role. Interviewee 94 was keen to stress he did not 'control' other press offices although the GMU took the lead when this had not been as clear-cut before. He argued, however, that with the GMU in charge, there was scope for individual press offices to shine in huge stories, with many member organisations in emergencies wanting their own offices to speak out.

> So if there's a Japan tsunami, if the Norwegians want to do something, or the Swedes want to do something, they'll do interviews out of Stockholm or Oslo in their own respective languages. From the perspective of [my old job at] the BBC, that's kind of weird—there, you'd want a voice out of Japan. But that helps me hugely because we have far fewer demands on key individuals who are in the field.

The effect of the GMU was to homogenise and control the Save the Children message, not just in one country but around the world. This meant that in a globalised and web-first environment, the aid agency's coverage did not contradict itself.

Conclusion

> The question is whether the international humanitarian community is willing to listen and put the voice of the survivors at the centre of their priorities and plans.
>
> (Gormley, 2014:83)

Despite the investment that NGOs are increasingly making in new media, many remain unclear about what they are doing or trying to achieve with it—symbolised by the fact that many were still working on social media policies. While in their policy documents, NGOs reiterate that they aim to make beneficiary voices heard, among the interviewees there was still uncertainty and nervousness about how best to achieve this most effectively.

It is true that there have been some imaginative efforts, as outlined in this chapter, at harnessing the power of digital media to overcome the distancing effect of traditional media, which often made the focus of humanitarian disasters donors or journalists, rather than survivors. Yet many aid agencies still see traditional media as the most effective way to spread their message—and insist they should remain the gatekeeper of stories from the field.

Save the Children's Global Media Unit was the most sophisticated way of trying to make sure the press office has control of messaging, not just across one country, but globally. The GMU aimed to secure harmonious messaging and the possibility of the same basic press release, tailored to different countries, being released simultaneously—helping to address some of the conflicting agendas of Save the Children US and Save the Children UK, for example. There had been conflicts, for example, over attitudes to adoption of orphans by celebrities such as Madonna (Hazlett, 2009), which made it into the media. Even after the creation of the GMU, this attempt to harmonise media messaging between different Save the Children organisations did not always work with internal stories. After the decision by Save the Children US to present an award for leadership in international development to former prime minister Tony Blair, more than 500 employees around the world signed a petition calling for its withdrawal (Rawstrone, 2014). Importantly, in the case of a rapid onset disaster, it allows different press offices around the world to ensure that Save the Children remains at the forefront of media coverage. That is because the delays involved in signing off press releases are minimised and press offices are told what has already been released to the media, as well as which media organisations have been approached. For the GMU there was little talk of beneficiaries having a voice in this process; the focus was on a 24-hour control of the message, co-ordinated from the powerful London offices of Save the Children UK, with the aim of reaching decision makers (politicians and business people).

Staff at many other agencies however were more willing to articulate the feeling that their gatekeeping role could not be sustained, and that they had a normative obligation to ensure more voices were heard. For many, though, handing over more control to beneficiaries was not seen as practical or even in the best interests of the agency.

One early trend, therefore, was to embrace the use of bloggers as a 'safe alternative'. While the bloggers were not beneficiaries, neither were they full-time journalists. Thus—like other forms of non-mainstream media—they were seen as more 'authentic'. Aid agencies often chose them because they have a large number of followers and are likely to blog and update on several occasions, if not every day, instead of just producing one final piece. The difficulties of a media trip with conventional journalists—at the mercy of a changing news agenda and the need for news pegs or case studies—appeared less intense with bloggers, for whom the story could be framed around their own personal experience.

Bloggers also tend to view being taken on a trip by an aid agency as a once-in-a-lifetime opportunity for which they feel immensely grateful. They frequently communicated this gratitude to their audience, rather than taking the 'critical friend' approach of most journalists. Waiki Harnais, for example, summed up her trip to Ethiopia with Plan UK as follows:

> Overall, everywhere we went, we noticed Plan had quite a strong presence in these communities, from health centres to schools and youth projects, all built by Plan and most handed over to the community. Although it was clear that the people in these communities still needed more help and support, it was very inspiring to see the difference that Plan is making in their lives. [. . .] I am hoping that you will be touched by some of the things you have read in this post, and will visit the Plan website to find out more about how you can help these communities.
>
> (Harnais, 2011 quoted in Cooper, 2015)

The waning power of the blogger and the growth of more visually led influencers such as vloggers or Instagram influencers meant that NGOs had to negotiate new boundaries—such as an increasingly media-literate community being able to seek out social media indiscretions from years back, and the problem around increasing commercialisation of social media influencers. As such they have tended to be cautious around the use of such individuals and returned to a more controllable media approach—whether using celebrities or producing their own social media content.

Notes

1. In particular, article 10 of the Code reads "In our information, publicity and advertising activities, we shall recognize disaster victims as dignified human

beings, not hopeless objects". See more at: www.ifrc.org/en/publications-and-reports/code-of-conduct/#sthash.Zc43Lkvq.dpuf.
2. The full Q&A can be read here: https://twitter.com/search?f=realtime&q=%23decqs&src=hash.
3. During DEC appeals, member agencies send press officers from their own offices to the central DEC office.
4. See: https://twitter.com/stuartfowkes/status/315070173168144384.
5. A BGAN is a global satellite internet network, which uses Skype. It has the advantage that it is the size of a laptop—i.e. far less bulky than previous satphones; a Quicklink allows high-quality compressed media to be sent via low-speed or unstable broadband connections.
6. A Storify (compilation) of Hasan's tweets and reactions to them can be found here: https://storify.com/oxfamgb/twitter-takeover-hasan-syrian-refugee.
7. Tearfund did take a non-white blogger on a trip to Cambodia in 2014.
8. Both men had left Save the Children in 2015, but Cox subsequently resigned from the charities More in Common and the Jo Cox Foundation, while Forsyth resigned from his new role as deputy director of Unicef after the complaints became public and Save the Children apologised to the women for the way it had been dealt with (Greenfield, 2018). In April 2018, the regulator the Charity Commission announced that it would be investigating the way the charity handled the harassment and misconduct claims (Perkins, 2018).
9. Save the Children UK used the term 'media moments' to mean a specific report/press release/stunt that would capture mass media interest.

References

Anderson, A. 1991 "Source strategies and environmental affairs" *Media, Culture & Society,* vol. 13, no. 4, pp 459–476.

Anderson, A. 1993, "Source-media relations: The production of the environmental agenda", in *The mass media and environmental issues*, ed. A. Hansen, Leicester University Press, Leicester, pp 51–68.

Anderson, A. 1997, *Media, culture and the environment*, UCL Press, London.

Banugra, S. & Powell, L. 2012, 16 November-last update, *Sierra Leone scraps tactile voting system for elections* [Homepage of The Guardian], [Online]. Available: www.theguardian.com/global-development/2012/nov/16/sierra-leone-tactile-voting-elections [2018, 9 May].

Bayoh, A. 2014, 15 October-last update, *Living in an Ebola crisis—'I'm just missing watching Everton and doing my Job'* [Homepage of Huffington Post], [Online]. Available: www.huffingtonpost.co.uk/amjata-bayoh-/living-in-an-ebola-crisis_b_5681909.html [2018, 9 May].

Becker, K. 2011, "Looking back: Ethics and aesthetics of non-professional photography", in *Amateur images and global news*, eds. K. Anden-Papadopoulos & M. Pantti, Intellect Books, Bristol, pp. 23–41.

Beckett, C. 2009, *NGOs as gatekeepers to 'Local Media': Networked news for developing countries*, LSE, London.

Benthall, J. 1993, *Disasters, relief and the media*, I B Tauris, London.

Bristow, J. 2005, 7 July-last update, *Who saved Birhan Woldu's life*? Available: www.spiked-online.com/newsite/article/886#.VMamf0soZg0 [2015, 26 January].

CARE International. 2013, *Care international evaluation policy* [Homepage of CARE Internationa], [Online]. Available: www.care.at/images/_care_2013/expert/pdf/Programming%20Principles/CI_EvaluationPolicy.pdf [2014, 17 December].

CARMA. 2006, January-last update, *The Carma report on Western media coverage of humanitarian disasters* [Homepage of CARMA International], [Online]. Available: www.imaging-famine.org/images/pdfs/carma_%20report.pdf [2012, 24 April].

Carr-Brown, J., Fielding, D. & Clifford, N. 2005, 3 July, "Embracing the face of famine", *Sunday Times*, London, p. 7 Home News.

CDAC Network. 2018, *Who we are* [Homepage of CDAC Network], [Online]. Available: www.cdacnetwork.org/about-the-network/who-we-are/ [2018, 9 May].

Chouliaraki, L. 2006, *The spectatorship of suffering*, Sage Publications, London.

Charityjob.co.uk, 2015 [Homepage of Charityjob] [Online]. *Head of Global Media Unit* Available: https://www.charityjob.co.uk/jobs/save-the-children/head-of-global-media-unit/395008 [2018, 26 August]

Christian Aid. 2018, *Listening and responding to our stakeholders* [Homepage of Christian Aid], [Online]. Available: www.christianaid.org.uk/about-us/accountability-and-transparency/listening-and-responding-our-stakeholders-accountability [2018, 9 May].

Clutterbuck, L. 2013, 27 February-last update, *Fulfilling a stereotype* [Homepage of Liz Clutterbuck], [Online]. Available: www.lizclutterbuck.com/2013/02/fulfillingastereotype/ [2014, 29 January].

Cooper, G. 2007, *Anyone here survived a wave, speak English and got a mobile? Aid agencies, the media and reporting disasters since the tsunami*, Nuffield College, Oxford.

Cooper, G. 2011, *From their own correspondent? New media and the changes in disaster coverage: Lessons to be learned*, Reuters Institute for the Study of Journalism, Oxford.

Cooper, G. 2015, "Unlocking the gate? How NGOs mediate the voices of the marginalised in a social media context", in *Media, margins and civic agency*, eds. E. Thorsen, D. Jackson, H. Savigny & J. Alexander, Palgrave Macmillan, London, pp. 29–42.

Cottle, S. 2000, "TV news, lay voices and the visualisation of environmental risks", in *Environmental risks and the media*, eds. B. Adam, S. Allan & C. Carter, Routledge, London, pp. 29–44.

Dogra, N. 2012, *Representations of global poverty: Aid, development and international NGOs*, I B Tauris, New York.

Ford, L. 2011, 20 August-last update, *Blog on lack of educational chances for girls in DRC wins plan UK competition* [Homepage of The Guardian], [Online]. Available: www.theguardian.com/global-development/poverty-matters/2011/aug/10/plan-uk-blog-competition-winner [2014, 17 December].

Franks, S. 2006, "The CARMA report: Western media coverage of humanitarian disasters", *The Political Quarterly*, vol. 77, no. 2, pp. 281–284.

George, J. 2010, *Beauty and horror: two sides to Bangladesh*, [Online].. Available: https://groups.yahoo.com/neo/groups/BdOsint/conversations/topics/7124?var [2018, 24 August].

Gormley, B. 2014, "The changing face of humanitarian reporting: Putting the survivor centre stage", in *The future of humanitarian reporting*, ed. G. Cooper, City University London, London, pp. 81–87.

Greenfield, P. 2018, 22 February-last update, *Unicef deputy quits after inappropriate behaviour claims* [Homepage of The Guardian], [Online]. Available: www.theguardian.com/society/2018/feb/22/unicef-deputy-justin-forsyth-quits-inappropriate-behaviour-claims [2018, 9 May].

Guardian Readers and Bannock, Caroline. 2014, 25 November-last update, *Ebola crisis: Sierra Leone citizen reporters share their stories* [Homepage of The Guardian], [Online]. Available: www.theguardian.com/world/guardianwitness-blog/2014/nov/25/ebola-crisis-we-meet-the-citizen-journalists-reporting-on-the-crisis [2018, 9 May].

Halstead, M. & Powell, L. 2017, "Tell it in your own words", *British Journalism Review*, vol. 28, no. 2, pp. 42–47.

Harnais, W. 2011, 27 November-last update, *A very personal encounter with Ethiopia* [Homepage of Waikiharnais.com], [Online]. Available: http://waikiharnais.com/2011/11/27/a-very-personal-encounter-with-ethiopia [2014, 5 November].

Hazlett, C. 2009, 31 March-last update, *Debate continues over Madonna's adoption* [Homepage of MSNBC], [Online]. Available: www.today.com/id/29961703/ns/today-today_entertainment/t/debate-continues-over-madonnas-adoption/#.VsscHksdUds [2016, 22 February].

Islamic Relief UK. 2018, *Get involved: The cake campaign* [Homepage of Islamic Relief UK], [Online]. Available: www.islamic-relief.org.uk/get-involved/the-cake-campaign/ [2018, 9 May].

Keoghan, E. 2010 1 September-last update *#Blogladesh: our visit to the children's wards* Nixdminx.com, Available: http://www.nixdminx.com/2010/09/01/blogladesh/ [2018, 24 August]

Klein, A. 2014, "Radar: From the margins to the front page", in *The future of humanitarian reporting: Working papers*, ed. G. Cooper, City University London, London, pp. 29–35.

Kortu, M. 2014, 4 July-last update, *How Ebola created a climate of fear in my region* [Homepage of Channel 4 News], [Online]. Available: www.channel4.com/news/ebola-virus-sierra-leone-kailahun-fear-moses-kortu [2018, 9 May].

Ma, A. 2017, 8 February-last update, *On our radar: 3 challenges of remote storytelling* [Homepage of News Rewired], [Online]. Available: www.newsrewired.com/2017/02/08/on-our-radar-remote-storytelling-challenges/ [2018, 9 May].

Manning, P. 2001, *News and news sources: A critical introduction*, Sage Publications, London.

Moeller, S.D. 2006, "Regarding the pain of others: Media, bias and the coverage of international disasters", *Journal of International Affairs*, vol. 59, no. 2, pp. 173–196.

Myles, P. 2016, 27 June-last update, *The story of Ebola from those who lived through it* [Homepage of New Internationalist], [Online]. Available: https://newint.org/taxonomy/term/6949 [2018, 9 May].

Nissenbaum, H. 2004, "Privacy as contextual integrity", *Washington Law Review*, vol. 79, no. 1, pp. 119–158.

Orgad, S. & Seu, B. 2014, "'Intimacy at a distance' in humanitarian communication", *Media, Culture and Society*, vol. 36, no. 7, pp. 916–934.

Perkins, A. 2018, 11 April-last update, *Save the children faces formal investigation over staff misconduct* [Homepage of The Guardian], [Online]. Available: www.theguardian.com/society/2018/apr/11/save-the-children-inquiry-staff-misconduct-brendan-cox [2018, 9 May].

Rawstrone, A. 2014, 22 December-last update, *Save the children US's award to Tony Blair puts Justin Forsyth on the spot* [Homepage of Third Sector], [Online]. Available: www.thirdsector.co.uk/save-children-uss-award-tony-blair-puts-justin-forsyth-spot/policy-and-politics/article/1326671 [2016, 22 February].

Save the Children UK. 2016, *Annual report 2016*, Save the Children UK, London.

Schlesinger, P. 1978, *Putting reality together*, Constable, London.

Schlesinger, P. & Tumber, H. 1994, *Reporting crime: The media politics of criminal justice*, Oxford University Press, Oxford.

Slim, H. 2011, October-last update, *International NGOs: A necessary good* [Homepage of Global Briefing], [Online]. Available: www.global-briefing.org/2011/10/a-necessary-good/ [2015, 26 August].

Van der Gaag, N. & Nash, C. 1987, *Images of Africa: The UK report*, Oxfam, Oxford.

Wardle, C. & Williams, A. 2008, "ugc@BBC:Understanding its impact on contributors, non-contributors and BBC news", *BBC*. Available: www.bbc.co.uk/blogs/knowledgeexchange/cardiffone.pdf [2018, 28 June].

6 Cloning and Co-opting

How Mainstream Media Control User-Generated Content

> The best reporters . . . can take their own video and cut it, they can take pictures, they can do voice interviews, they can write like a dream, they can tweet their locations and interesting stuff to gather people to follow them, and they can do occasional blogs and put in pictures. I think that's your ideal reporter, somebody who can do all those things. Now there are very few people who can.
>
> (Interviewee 21, foreign editor, broadsheet)

Interviewee 21's perfect journalist is clearly a multi-tasker *par excellence*; but in only one short sub-clause does the foreign editor of this broadsheet newspaper mention that this dream employee might actually have to be able to write a news report for a living. Instead, he focuses on how the journalist can absorb the conventions of user-generated content in order to fulfil their duties in the early 21st century.

User-generated content (UGC) was initially defined by how it differs from traditional journalistic normative values—its focus on the subjective, connective and raw, with different boundaries of taste and distinctly different news values (Allan, 2006, 2009, 2013). However, more recently, researchers have turned their attention to the convergence of these values as eyewitnesses ape traditional journalistic norms to bolster their own credibility and journalists employ some of the tropes of eyewitness media—the first person, immediacy, lack of editing, publishing first by social media—that focus groups say they find attractive about UGC (Wardle and Williams, 2008). Some saw them as indistinguishable: as another broadsheet foreign editor commented, "Blogging isn't any different from reporting, really" (Interviewee 29).

This chapter looks at the so-called 'Gupta effect'[1]: mainstream journalism's use of UGC techniques to absorb challenges and defend itself against incursions. By cloning the very strengths of UGC, journalists can regain some of the ground they have lost to it. The chapter also looks at two specific examples of influential media organisations—in this case the *Guardian* and the BBC—using different approaches to control and co-opt UGC as a way of protecting their own journalistic authority.

This chapter is based on time spent researching at the *Guardian*'s newly formed GuardianWitness platform and the BBC's UGC hub in 2013.

Moving From the Citizen Journalist to the Journalist Citizen

Conventional reporting of humanitarian disasters in the past worked to a template: the dramatic scenes of flood, quake or famine; the call to action (aid needed *now* to help the victims); and the quick results—heroic aid worker rushing to the scene, food parcels arriving, doctors setting up temporary medical care. But, crucially, the journalist was at the centre of it all, explaining what was happening. One of the most highly prized normative values in journalism is that of the act of witness—the journalist as truth-teller, exposing the untold story because s/he is there to see it for themselves, as a reliable, trustworthy witness, watching the crisis as it unfolds:

> Journalists guarantee the veracity of events and they provide the link between the world of the comfortable and the familiar and that of the victim. More recently—only over the last decade—victims have been invited more frequently to speak for themselves.
>
> (Seaton, 2005:188)

The authority of presence, however, is one that can prove fraught with difficulties because of the tensions of interpretation and memory (Allan, 2013; Kitch, 2008). To balance the urgency of first-person testimony and the emotions it may unleash, journalism has commonly employed the codified values of impartiality, offsetting the personal with the professional ethos of distance and detachment.

This has always been a difficult tightrope to walk, particularly in the potentially emotive area of reporting humanitarian crises. Like those who cover war such as Bell (1998) and his 'journalism of attachment', journalists who find themselves face-to-face with destroyed lives can become activists and witnesses, who charge themselves with keeping the faith. They take what appears is an unassailable position: that it is bad to let people starve to death or lose their homes.

This is not a new debate; what differs now is the way that UGC is affecting how disaster coverage is framed. First, the personal has become ever more privileged. Allan characterises citizen journalism as a "type of first-person reportage . . . in order to participate in newsmaking, often spontaneously during a time of crisis, accident, tragedy or disaster when they happen to be present on the scene" (Allan, 2013:9).

This content is based on connection rather than fact (Matheson, 2004). Journalists no longer saw themselves as bound to one news programme, but needed to interact with their audience constantly—cloning the methods that their viewers and readers employed. This primarily involved tweeting either messages, photos via Facebook and Instagram—and these days using livestreaming services such as Facebook Live or platforms such as Snapchat Discover—much as the audience would share with their friends. This placed the journalist right there at the scene, without the audience having to wait for a news bulletin or piece on the website/newspaper.

Conversely an international editor for a commercial broadcaster said that people had contacted him on Twitter to direct him to UGC of people caught up in humanitarian crises.

> So there is, between the journalist and the afflicted community, an interaction now that there never was before, and I suppose it empowers people who at that point are powerless, and you know, alerts people like us as well as potentially authorities they're in difficulty, or that something is unusual in their area.
>
> (Interviewee 38)

This development was not welcomed by all journalists though. A foreign correspondent on a quality newspaper thought the paper's first-person accounts were less considered because journalists no longer had the time and space to sit down and write a piece:

> The whole thing with the web, there's now such a demand for upbeat stories or we've got to be across the news . . . so spending time with people as opposed to talking to them for 20 minutes then moving on [has gone]. What we used to do is I'd go and spend two or three hours with somebody just sitting there, talking to them, really getting under there, talking to other people in their village, and writing a whole thing about my experience. There's less time for that, and more demand to file and update.
>
> (Interviewee 30)

Interviewee 27, a foreign editor on a Sunday broadsheet, also saw dangers in UGC-style privileging of the personal.

> It's all very well but it doesn't tell you what the overall situation is, what the numbers are, how it's going, what is being put in place and so on. So, it is often shallow and knee-jerk and fulfilling a sort of pre-arranged script that is handed down from disaster to disaster by news desks and editors.

Social media also killed the idea of the exclusively broadcast or print reporter—and the notion of one daily deadline. Sunday newspaper reporters still have relative freedom, but those on dailies or in broadcast journalism are increasingly expected to work in as many different media as possible, so that their organisation is seen to be on top of the news agenda. An assistant editor in newsgathering at a commercial broadcaster summed it up as follows:

> If someone arrived on the scene of an earthquake at 2pm, and I want a piece by six or ten [o'clock], I probably wouldn't labour them with requests for new media offerings. But there's various things you can do, you can be in a car on your way to the earthquake and take a picture out the window, send it back if you've got a mobile phone signal, you can send a few lines . . . we'd encourage them to do that.
>
> (Interviewee 37)

For journalists, Twitter remained the most important social medium, as both a source and means of disseminating information—and to show their readers or viewers they were on the ground.

This was not just the case for broadcasters, but for newspapers as well. A foreign editor at a broadsheet paper described how tweeting on the ground could increase traffic to the website. In his previous job at another broadsheet, he recalled that a reporter who was first on the ground in Libya tripled his Twitter followers within 24 hours because he was seen as an authentic source on the ground—and in the process generated more traffic for the newspaper's website. Increasing Twitter followers to increase web traffic has become policy elsewhere.

> The big thing is if you're being followed, or if you're in a particular position of one would say privilege when it comes to a news story, then people will follow you, and by that right they will also follow the stories that you write, which means that the traffic that is generated to the website will be better.
>
> (Interviewee 21)

There were, however, some small pockets of resistance to the employment of such methods; many media outlets thought the 'digital first' approach—especially tweeting snippets of information—risked costing them their edge and unique information. It is perhaps unsurprising that Sunday papers—which only have one chance a week to make an impact—were the most opposed to it. One Sunday foreign editor took the view that reporters tweeting potentially caused more problems than it was worth—that stories would be given away. A foreign editor for a broadsheet was the only journalist who thought reporters tweeting and taking photographs could actually enhance the *story*, rather than simply promote the media *brand*.

> I don't remember [our reporter sent to Haiti] tweeting a lot but he did take his own pictures. Not front page pictures, but they were really helpful so you knew the person he was talking about. He wrote a piece about a wheelbarrow of bodies he'd followed and it wouldn't have had anything like the same power without that picture.
>
> (Interviewee 24)

What many editors were beginning to realise, however, was that there were risks to reporters who used social media from the ground. Caught up in the middle of a dangerous humanitarian crisis or a war zone, editors sometimes had to warn their staff that they were potentially putting themselves in jeopardy.

> I've had to stop one person from tweeting, for example, 'X, who was in Y', and referring to it as the kidnap capital of Syria, and tweeting that she was there, thus alerting any number of undesirable people to her person, at one of a limited number of hotels. She has only 4,000 followers, she's addicted to tweeting, she felt that it connects her with people, when she's the only Western journalist in a difficult part of Syria, and I can understand that, but I had to tell her that her security was more important and she just had to stop.
>
> (Interviewee 20, broadsheet foreign editor)

Interviewee 21 also felt that, in "dodgy" situations, journalists should not be advertising their presence—no matter how attractive the idea of promoting themselves as being in the centre of the action. For his rival on another broadsheet, however, the main problem was getting the more senior reporters to engage in social media at all because they saw it as beneath their dignity. This task was doubly difficult because, as he put it: "It's very hard if someone is risking their life to be stern about tweeting" (Interviewee 24).

Research in Norwegian newsrooms (Brandtzaeg and Chaparro Dominguez, 2018) has found that even when senior journalists (aged 42–55) could be persuaded to take part in social media, they tended to stick to 'elite' networks such as Twitter while younger journalists (aged 21–26) were prepared to embrace Facebook, Twitter, Instagram, blogs, YouTube, LinkedIn and Snapchat. What was particularly interesting was that while older journalists privileged Twitter, the younger journalists saw it as an information bubble that excluded the general population and thus needed to go beyond that when searching out sources on breaking news stories.

For those on the ground the idea of transforming themselves into a multimedia journalist was taxing. An international editor[2] for a flagship bulletin, while echoing her editor's idea that the main news bulletin was key, was also having to produce all sorts of different content:

> We are online, so we're expected to constantly produce blogs and tweets, and the cameraman I worked with in Mali does films as well, and photo galleries. We have one major focus which is the 7pm programme, and that should never be sacrificed to anything else. But there is a demand for online as well, which we try and fulfil as much as possible . . . 100 per cent is devoted to the seven o'clock bulletin, really, and the extra 20 per cent that you put in is devoted to the other bits, and I write for the *Sunday Times* and the *Observer* as well, and do Irish radio, television, anything.
>
> (Interviewee 36)

A foreign affairs editor for a satellite broadcaster said that one of the steepest learning curves for reporters had been coming to realise they could no longer regard an organisation like his as a 'TV news organisation'; they had to see it as a multimedia organisation with reporters able to turn their hand to any form of communication (Interviewee 40). While there was still the budget to send TV reporters to cover humanitarian disasters, there was no money to send online or radio reporters; the TV reporter now had to fulfil those roles as well—writing a 500-word lead for the website and a 30-second radio track.

Some reporters felt that the pressure to perform these extra tasks was taking them away from more important journalistic work. One reporter who had covered emergencies including the 2011 Great East Japan quake and Typhoon Haiyan in 2013 for the broadsheet he worked for, was clear: there were limits to what the editors back in the UK could expect.

> I mean I filed a couple of days for the web, but essentially when I got to [Tacloban, in the aftermath of Typhoon Haiyan] it was pretty hard to file. There was no electricity, it was pretty tough, so they got one piece a day, I guess.
>
> (Interviewee 25)

Interviewee 40 said his editors had some understanding about this new pressure, but it was still intense. "There used to be occasional downtime, there really isn't any downtime any more".

The BBC's UGC Hub: Boundaries as Shared Spaces

The BBC was one of the first media organisations in the UK to realise the potential of user-generated content and attempt to harness it. The BBC's UGC hub had begun as a 24/7 service in 2005—a week after the 7/7 bombings in London. The attack proved the catalyst: the corporation received 1,000 stills and videos, 3,000 texts and 20,000 emails from the public within 24 hours of it taking place (Boaden, 2008); the scale and quality of the public's contributions made the BBC, in the words of Richard Sambrook, former director at BBC Global News, "cross a Rubicon" (Belair-Gagnon, 2015). Two years later, the Hub moved from its small office on the upper floors of TV Centre to the BBC's main multimedia newsroom, signalling its increasing importance; by 2008 the corporation had updated its editorial guidance to journalists to include social networking. Events such as the Saffron Revolution in Burma in 2007, the 2008 Mumbai attacks, the 2009 Iranian elections and the Haitian earthquake of 2010 put user-generated content on the map, and social media helped establish the hub as crucial to newsgathering (Belair-Gagnon, 2015).

The Hub has proved a popular place for researchers to study. This is down to the BBC's relatively open attitude to scrutiny and the comparatively large pool of journalists it has assigned both to looking at content sent in by the public and to soliciting such material. Harrison (2010) defined four types of UGC found at the BBC: UGC as a form of unsolicited news story; UGC as a form of solicited content for specific extant news stories; UGC as a form of audience-watchdog content; and UGC as a form of expeditious content for specific items and features in the future.

The first two are the most relevant, although the last two also have their place (for example, the audience could at times be critical of the BBC's focus during disaster stories and UGC journalists at the BBC would plan ahead for disaster 'anniversary' pieces). The Hub was also under pressure to stay ahead of its rivals, many of whom were enthusiastically using UGC and accessing it via the wire agencies or agencies like Storyful.

While Bourdieu (1998), Gieryn (1983, 1999) and Abbott (2014) emphasise the struggles around demarcation of social space and the creation of boundaries as a way of emphasising journalists' cultural capital, it is worth considering how boundaries can act as "means of communication as opposed to division" (Lamont and Molnar, 2002:177). In particular, Star and Griesemer (1989) suggest that heterogeneous groups can co-ordinate through the concept of 'boundary objects'—objects that let different groups protect their identity while allowing for shared communication. Star later defined the boundary object as the 'shared space' (2010).

The BBC was an interesting exposition of this idea. By setting up a separate organisation—the UGC hub—it gave UGC a status that other media organisations did not. The journalists there felt that people turned to the BBC, as the national broadcaster, at times of crisis and gave it their content as part of a collaborative process, in a way that they would not do with other media organisations. The BBC reflected this collaboration by making UGC part of all areas of the BBC—TV, radio and online.

And yet there was still a strong sense the UGC journalists synthesised and controlled the content. While UGC was valued, it was still treated very much as a starting point. It was to be used as raw material—the basis for harvesting case studies and gathering eyewitnesses, whose words could then be moderated and translated through the medium of the journalistic interview. As one UGC producer, put it:

> Producers are quite lazy and they will ring the Hub for case studies, and I will then say "have a look at the comments widget [on the website] and if there's any you like, give me the name and I can go and pull out their email address. But I don't have time to go through 400 comments and find some that might suit you".
>
> (Interviewee 83)

Set Up and Purpose of the Hub

In 2013, a team of around 20 journalists worked a shift pattern at the Hub, with the main action happening first thing in the morning. The overnight producer gave a handover note to the early producers who came in around 6am, so any breaking stories, particularly from the US and Asia, were covered. The early producers then concentrated on working through likely content for morning programmes. They looked at the main stories on the news website as the primary vehicle for generating content, and then decided whether there were any stories that might need Hub input in terms of pursuing audience contact or content.

On a typical day, the assistant editor came in at 8am for a succession of editorial meetings (8am, 8.30am, 8.50am and 9.30am) covering world news, domestic online news and general newsgathering. These meetings examined what the other BBC news teams were planning for the day, before the Hub's own editorial meeting at 10am, when the team would go through the running order for news output. From that point onwards, the assistant editor ensured producers were assigned to stories, deals with specific requests for UGC from programmes, stayed across the newswires and manages referrals. There were also two editorial meetings later in the day (3.15pm and 3.30pm) for forward planning for the next day.

> Our core functionality is to try and get pictures, videos, real-life voices on the website as quickly as possible and to enhance what our coverage is already doing, so there's no point for us to replicate stuff that's already happening elsewhere. I guess our function is an internal sort of news agency. Every bit of BBC News is effectively our client which we try to push as much of our content or material to, and try and get it on air or published on the website, and so we're very much there to help the rest of BBC News.
>
> (Interviewee 82, assistant editor, UGC Hub)

Hub reporters would also produce first-person ghosted pieces called 'Voices', writing up verbatim an interview with someone caught up in a news story. These were not read back to the subject unless particularly sensitive ("We should be trusted enough, we should be good enough at our jobs to be able to do it", said one UGC producer), nor were they given bylines by BBC staff.

The Hub worked with an outside moderating firm called Tempero[3] to facilitate comments on the website. A widget was put on the bottom of certain news stories, often at the request of other programmes or after discussion in the Hub as to what people would be most interested in. No more than six widgets were put up per day, and they would go on a mixture of national and international stories. A duty producer would then select the 'editor's picks', which would be the stories rated highest by readers, and ensure the further post-moderation of any off-topic or offensive comments, in case it had been missed by Tempero. Sensitive stories were pre-moderated, which meant that every comment had to be read and approved before going up on the website. The UGC producer said that the most popular stories (Prince Harry photographed naked in Las Vegas was one example she cited) would run to 1,500–2000 comments before the widget was closed.

The main function of the UGC Hub at that time, however, appeared to be a technically sophisticated way of finding 'case studies'[4] for different parts of the corporation. A senior broadcast journalist flagged up 'editor's picks' from comments that were added to webpages, for example, as people who could be contacted, as well as accounts via post forms, which could be passed on to radio and television programmes. This led her to describe the Hub as "reactive" rather than proactive, although on occasions they would put up 'calls to action'—forms requesting people's thoughts or memories to anticipate forthcoming newsworthy reports or anniversaries. Sometimes, however, this was also about driving traffic to the BBC news website. "It's about putting stuff out there to get people to click back and view the website" (Interviewee 84).

She saw emails as the most efficient way of collecting case studies. Texts frequently left out names and phone numbers, and there was no easy way of geolocating the sender. Twitter, however, was becoming increasingly important, while Facebook came into its own after dramatic events such as a murder:

> Say with the Lee Rigby murder [the murder of an off-duty soldier in London in 2013], the tribute pages are quite helpful. Also, with Facebook, things that can be linked in as well, for verification purposes. So it might be that someone said Drummer Lee Rigby (before he was officially named) or "we think he might be from Rochester and we think he might have one kid or something". With Linked-In or Facebook you can piece together the jigsaw, and the more you can either confirm, "Yeah it's more likely to be him" or actually "no, cos this guy's got five kids or is 42" so that's quite helpful as well. Not necessarily to get people to speak, but to try and piece together the puzzle.
>
> (Interviewee 84, senior broadcast journalist)

The assistant editor said the Hub was so multifaceted that one part of the BBC that used its services might well have no idea what it offered to other departments. So, for example, the World News Channel might interview Hub journalists on how social media was playing out on a big story, or how the BBC was checking content. Other parts of the BBC news operation would regard the Hub's regular briefings on audience sentiment on a day's news topic at the 9am or mid-afternoon news conferences as its most valuable contribution. While for other teams the Hub's content database for an upcoming story was what they most wanted to raid.

> It's a bit like if you're a blind person and you're trying to work out what an elephant looks like. You sort of see one bit of the operation and you think that's all it will affect, that's what their role is . . . they don't really know what else we do, other than that one thing.
>
> (Interviewee 82, assistant editor)

When a humanitarian story broke, in the words of one UGC producer, who was then acting as Live & Social producer which meant sitting in the main newsgathering hub to 'feed in' important developments from UGC to television, radio and online, "anything you can find has value, because no one has anything" (Interviewee 87). So tweets, stills and video all played a crucial role in helping to tell the story in those first hours. Social media monitoring would focus on those sites that can easily geolocate users, because people at the centre of the storm, earthquake or wave often have a network of contacts who share information with them, so they effectively become a filter of content.

> That's a role we've seen happen a lot now with breaking news stories, where people, just for a temporary period, become great news stories just because of their expertise, their local expertise or their contacts and they may never do that again. . . . So trying to find those sorts of people is pretty critical. . . . They may be able to help us find other people, other sources, help us verify or credibility-check the content.
>
> (Interviewee 82, assistant editor)

Interviewee 82 said the BBC were trying to develop an algorithm to identify such accounts based on previous activity or connections, but it was obviously a difficult thing to achieve with rapid onset humanitarian disasters. They had enjoyed some success with longer-running stories, in identifying well-placed Tweeters or YouTubers by monitoring and crosschecking what they were saying against official sources.

Interviewee 84 said that with a big story like an earthquake different producers would be tasked with monitoring Twitter, Geofeedia, the Have Your Say forms and the console on which pictures sent in directly to the BBC by users would arrive. This console was tightly password-controlled, with only a selected list of users "in the dozens not hundreds" able to access it.

Because parts of the world were still internet "black zones"—for example, remote areas of Asia and Africa—there was a struggle to find "real voices on the ground" during some disasters. As a result, most of the intelligence tended to come from expats who watched BBC news programmes abroad.

> They tend to be the first eyewitnesses we get on air, and then they also become great sources for us in terms of their networks, the people on the ground, their friends, family or other people that they suggest we should try and chase up.
>
> (Interviewee 82, assistant editor)

Those who contacted the UGC Hub tended to represent a similar demographic as the BBC's audience as a whole, but consisted of a wider range of people than might be immediately apparent.

> I guess internationally there's more of a bias towards English-speaking people and certain countries are better than others so, North America, we get a very big response, South America increasingly is getting better at that, there are some black spots in Europe, French people generally never bother contacting us and then Africa, we always get emails from Nigerians about every story, except stories about Nigeria, when we actually want them to get in touch.
>
> (Interviewee 82)

The Hub was building closer links with the BBC's different language services because of the increasing interactivity that those services had with their audiences. The BBC's Persian and Arabic services, for example, had almost created their own 'mini-Hubs' in order to find and verify UGC voices and content. So, if there was a flood in Mozambique, Interviewee 82 said the Hub would work closely with the BBC Arabic service in Africa to see which pictures, speakers and videos they had and ensure that their content was pushed out to the appropriate output teams—such as the News Channel and World News.

There was also a tension between the voices the BBC felt it should be using and the voices the Hub could provide. One UGC producer pointed out that there was a "British bias" on the site and while, for example, there was a strong feeling at the BBC during the Great East Japan quake that there should be more Japanese people quoted, the journalists said that the audience wanted to see people with whom they identified (Interviewee 86).

The use of Twitter and Skype during the Haitian quake changed disaster stories, according to a senior broadcast journalist at the Hub. She said that the corporation got the Have Your Say post form up rapidly, but there was "nothing useful" from that—only a very self-selecting group of expats who were "BBC news people". Twitter, on the other hand, was extremely useful.

> I think we had three [Twitter] accounts, three people that we were following, who were on the ground, and their tweets were amazing. So it was, "just woken up, what's happened, felt the tremor, I'm going to leave the children here, I'm going to go and see what's happened", and constantly tweeting about, "oh my God, never seen such devastation", and it was just fascinating. . . . At that time I don't think that we were quite aware of how important social media was, because these were the only first-hand accounts that we had.
>
> (Interviewee 88)

The assistant editor had described the post forms in general as "elegant" and useful because they were only seen by people who had read or scrolled through to the end of a story; this filtered out a lot of perhaps less informed voices. But Interviewee 88 said 70–80 percent of the people filling in the post forms were expressing sympathy, 20 percent were "useful" (for example, people with relatives caught up in the disaster zone) and no more than 5–10 percent were what she called "the real wow"—those who were on the ground with a story to tell or visuals to share.

A step forward was combining the use of social media with Skype. Interviewee 88 said that the UGC Hub would identify potential case studies—such as Carl Pedre, a Haitian radio show host who was tweeting about the quake—and not only use their content, but interviewee them too. Pedre could Skype because a generator at his radio station kept his internet connection going.

> That was massive. Before that, Skype was laughed at and then it was "oh my goodness, we can actually speak to somebody in Haiti". I think the audience are prepared to tolerate poor quality to get something in vision from the place where the story has happened.

But the BBC was facing a challenge. As Interviewee 87, an UGC producer, pointed out, many people were turning to social networking sites, which were growing rapidly, rather than the BBC to share content. There were still specific groups and situations[5] that turned first to the BBC but from around 2011 onwards pictures and videos tended not to be sourced through direct contact with the BBC, but through social media—in particular Twitter, Facebook and YouTube.

Permission and Verification

The use of social media, however, had changed in the time that a senior broadcast journalist had worked at the Hub. She found increasing suspicion and reluctance among UGC creators to let the corporation use their pictures. She put this down to two factors. First, was a sense that the BBC had not safeguarded content in the past:

> That really has changed. That's something to keep an eye out for, really. When the July 7th bombings happened [in 2005], you know, the guy with the picture of the bus with the roof ripped off, he sent that to the BBC, the *Daily Mail* then nicked it, and he was furious. He said: "I sent it to the BBC because I trust the BBC and I want them to have it for free, I don't want it just to get ripped off".
>
> (Interviewee 88)

Second, she thought creators had a stronger inclination to monetise such content, often egged on by other tweeters, and the BBC would only pay for photographs or videos if the content was "amazing". Interviewee 86 talked about payment being a "quagmire": "you just don't pay, and if they ask about payment, it's not our decision, it goes higher up". So some creators tended to tout their content elsewhere.

For those who did co-operate with the BBC, Hub journalists would send creators a set of terms and conditions that amounted to "you don't lose copyright but you lend the BBC the right to use it", according to Interviewee 87.

> Basically we have a conversation with them and say "are you happy for us to use it?" and then we'll send [them] the terms and conditions. Usually people are happy to use it and if they say they want payment, we'll send it to somebody else [higher up]. The point is that some agencies and some news organisations probably don't even contact people, they just use it and . . . might put YouTube or the name of the person and YouTube.

A senior broadcast journalist confirmed that BBC policy was to check whether content creators wanted to be credited using their username or their actual name. Some users preferred their Twitter or YouTube username, because that promoted their account. Any content sent out by the Hub had the preferred name on it.

> Copyright is something we're quite hot on. When we get stills verified, and video, we'll always say you must credit the name. Online's not too bad, but in TV they're quite reticent to have loads of astons[6] popping up, especially on something like an amalgamation of a few different stills. So that's difficult for us because if we're trying to get permission to use some really good footage, because we don't generally pay, we'll say you get a name credit.
>
> (Interviewee 84)

The assistant editor said that the Hub referred requests for payment to the main news desk—and the Hub also had to check with the creators that they were happy for the BBC to use their content across all outlets—online, TV, radio and language services—when they might not want it to have that reach. Interviewee 86 said that this often meant withstanding pressure from editors around the BBC eager to use a particular picture or film.

> One of the things I got asked in my interview when I applied to work in the UGC Hub is what would you do if an editor came up to you and wanted to use a picture straight away, would you be scared to stand up? You learn to be bolshy.

The assistant editor said that there were different levels of permission to go through. Ideally, a Hub journalist would get hold of the person who took a video, speak to them directly and get direct authorisation. Sometimes this wasn't possible, however, because of technical or safety reasons. In Syria, for example, the person who uploaded a video to YouTube was often not the person who had shot it; for security reasons the individual doing the filming didn't want to reveal their identity. If other checks meant the Hub was confident the material was credible, then it would pass it on to news programmes. The Hub would instruct the programme to use some cautionary wording around the content's use, so that the audience was aware that it had not been verified by talking to its creator.

Photos, however, were trickier—the BBC was not covered for 'fair use' of photographs, according to Interviewee 82.

> There will still be times when news programmes may still decide to run pictures, so all we can do is advise them that the picture is genuine because of x, y and z check that we've made, but we've not been able to get copyright clearance to use it. It will then be down to individual editors to take on the risk to broadcast it if they think it's sufficiently of value. So for example the Woolwich attack, there were some stills that we started running on air before we got official permission from some of the people who had taken them, so we'd already mentioned to them on Twitter, but we were still waiting for them to reply. In those instances it's less of a risk because you'd see that they'd already replied to two or three other news organisations to say yes, which doesn't clear us for using it but there's a good chance that they might once they get round to our message.

The Hub was seen as a centre of expertise for verifying user-generated content. Journalists said that the reputation of the BBC as a trusted source was such that they had to be extra careful. This meant that, as well as verifying content they came across by themselves, they would have to verify content sent in to the corporation.

> You get hoaxers. You've definitely got people who will lie, wasting our time and theirs. You have to phone them, you have to check, "what is it like where you are?" You check the weather, you check what time it is. This has happened to me—people put down their friend's phone number as a joke, and they're "oh, what are you ringing me for?"
>
> (Interviewee 86, UGC producer)

There was a lower level of verification for reproducing text such as tweets. Her view was that "at the end of the day, they're engaging in conversation so we will put it on there . . . I haven't had any complaints". However, if the tweet contained a picture then the verification procedures kicked in; she said the BBC had been "burned" before when people had retweeted pictures or taken other people's pictures.

Like many journalists, she believed strongly, however, that her journalistic instinct would alert her to potential problems with verification:

> I trust my gut feelings, and I will go round and I will check things. I think the more you do it, the more you do trust your gut feeling, the more experienced you are, so you know when something's a bit dodgy.
>
> (Interviewee 86)

Another UGC producer, Interviewee 87 echoed the need to talk to content creators and described how the Hub would try to track down creators who were difficult to find by matching their photos with Google Street View. The problem was that after devastating disasters like earthquakes, "there's no landmarks any more, it's just rubble". If, however, they could establish an image was from the right area, they could attempt to check its originality by doing a reverse image search on Google Images or TinEye.[7]

Duty of Care to UGC Creators

When a big story broke there was pressure on the Hub from the BBC's flagship programmes to speak to eyewitnesses. Interviewee 86 said it was at this point that the Hub's journalists needed to be most sensitive to the needs of the creators of content because often they "hadn't had time to process anything . . . they're literally there outside the school or in the middle of the earthquake". Producers had to try to ensure that the same content creator was not contacted by several different people. They also needed to identify whether someone who might have been suitable to speak to a website journalist might not have been ready for the glare of live TV broadcasting.

She said that the other danger was that people who had spoken to such a high-profile broadcaster as the BBC—often before they'd had time to process fully what they'd witnessed—ended up being trolled or abused on social media.

> You are literally unleashing a monster. You have no idea what the response is going to be, especially, and it sounds horrible to say that, because I am on the monster side. But people take things out of context, people start abusing, you get trolls . . . it's a nightmare situation to be in.

Interviewee 88 also raised the question about a duty of care to creators of content. In her view, the BBC and other journalists were not always treating creators as respectfully as they should, especially in extreme situations; she cited the example of an eyewitness to the Mumbai shootings talking to the corporation from under a table in the restaurant.

> You have to be very respectful, it's no different really from knocking on somebody's door. You don't steam in and say "can we get on air in five minutes?" You talk to people, you have an awareness of what they've seen. You also have a duty of care to them if they do come on air: how are they treated? What happens if they suddenly break down mid-interview? That has happened. What happens if they're phoning and they're being shot at, like in the Mumbai shootings? A guy lost his life, one of our eyewitnesses. Some journalists have got the pressure of an editor saying "just get that person". They will go in and can actually be quite damaging to the relationship we have with these media sites.

Failing to treat creators with respect could be bad for the corporation. Interviewee 88 said she had seen sites like Twitter "turn feral" on the BBC, as a result of what people perceived as insulting behaviour by journalists.

Graphic Imagery

Interviewee 87 said it was fairly "obvious" that the BBC wouldn't show "mangled bodies". Its correspondents would occasionally verify videos showing people dying so they could talk about them as part of a report, without showing them directly to the public (or they'd show only the bits of the video that didn't contain scenes of death). In some cases, even non-graphic images would be avoided.

> I guess you would want to avoid where you see someone dying, or you don't see them dying, you'd see the moment when death would happen, so there was a balloon crash in Turkey not long ago, and there was a video we had which showed the balloon hit the earth, and they didn't want to show it because that was the point at which people died, even though you don't see anyone die.

He said he thought there was "a BBC policy of some kind" prohibiting showing the moment of death, although he couldn't say for certain. What he definitely noticed was the BBC's attempts to reassess how it used graphic images on TV and online, both because of the changing attitudes of other broadcasters and in light of developments during the Arab Spring. The death of Muammar Gaddafi in Sirte in 2011 was a particular case in point:

> What the BBC did was controversial, because we showed online pictures of his dead body quite soon afterwards, and got a lot of complaints about it. Online struggled with that. Showing it on TV was one thing, it just flashed up and disappeared, but having a still of his mangled face, his bloodied face, was quite strong, on a site, which is easily accessible to anyone.

Dubberley, Griffin and Bal's (2015) work illuminated the potential problem of journalists being exposed to trauma through watching so much UGC of distressing events. This was something that Hub producers brought up time and again—the occupational hazard of seeing extremely distressing videos and stills without warning.

Interviewee 84 pointed out that it could be a shock for producers who were searching through the news and spam sent in on the picture console to suddenly see dead bodies or traumatic sights.

> If you're on the picture console you could be looking at pictures of chairs or themed things like flowers then suddenly you might see a

> decapitated body. . . . We had some of a guy who looked like he had been run over by massive tyre tracks and that came in about eight times from different accounts.

One producer, who did not want to be named, said that s/he had even delayed trying for a baby for a year because of a particularly horrific video s/he had seen of a three-year-old girl.

> When you're in the field, you know you're away from home, you've been prepared for it [traumatic sights], but I think here, I've got my Starbucks in my hand, I kiss the children goodbye, I come into work and I see mutilated children, dead bodies. My [spouse] and I joke about "how many dead people did you see today?" If I see less than ten it's a good day.

Another said s/he was still struggling to work in the department because of the images s/he had seen—not so much from Gaza because they "were always horrendous", or Syria because "we've become immune or used to it"—but from the one-off, sudden disaster stories.

Interviewee 86 said s/he had had to get an 'attachment'—a BBC term for a temporary assignment to another department—after two years of watching distressing stories involving babies, "especially when I had a baby in the house".

> It was too much. Not that I couldn't cope, I am a good journalist, but there comes a point when this is a lot to take on board and I think you have to be in a certain frame of mind and I got to the point when I was like I actually need to have a break from this.

S/he mentioned that they often worked with the Syrian and Libyan teams who had been warned that such videos could feature family members. She also worked closely with BBC Monitoring, who advised the Hub about how to deal with watching graphic content—ensuring that they took regular breaks, for example. The problem with UGC was that there were often no warnings on the video; the Hub started using warnings when sending distressing content on.

> Yeah, it's nasty stuff. You're a journalist, you're trained that it's not your grief . . . but you can't sit there practically every day watching children being blown up or being cut in places that really no-one should ever be cut, or just people grieving. As a human being you are affected, and I think people have different ways of coping. Personally, especially with the fact we do shifts, if you're there by yourself at night watching it, it's not nice. . . . In a way, yeah, it's toughened me up, but I do think personally people should be allowed, should be given a break from it.

Interviewee 87 said that the Hub worked hard to ensure that the same person was not doing 'the Syria shift' for a week. BBC Monitoring had also suggested other strategies to avoid distress, such as minimising the window of time during which the producers watched upsetting content and viewing it with the sound down.

GuardianWitness: Community or Control?

In contrast to the BBC, which saw UGC as something to be spread across all departments, the *Guardian*, with its GuardianWitness app, sought to segregate it (Wahl-Jorgensen, 2015) by establishing a particular 'site' for UGC. It was also—by actively soliciting both non-news and news content—aiming to co-opt participation by audiences in a controllable way, although the journalists themselves saw their work as 'ethical' and providing new spaces for diverse voices to be heard.

The *Guardian*, a left-leaning 'quality' UK newspaper, has invested heavily in its online presence and pioneered 'open journalism', a participatory and two-way process.[8] According to the then social and communities editor of Guardian Online, GuardianWitness was born out of a week dedicated to innovation at the *Guardian*, when various ideas were considered, including a demo called 'Guardian Contribute'.

This demo caught the eye of the then head of digital strategy who had been talking to the commercial partnerships team about doing something with *n0tice*, the hyperlocal blogging company. Discussion led to the idea of an app being set up, and the *Guardian* approached mobile phone company EE with the basics of the idea that would later become GuardianWitness.

GuardianWitness was situated within the community team at the *Guardian*. In 2013, when I was there, the team consisted of around ten content moderators and ten community co-ordinators assigned or 'embedded' into particular desks to ensure different parts of the *Guardian* were committed to open journalism. A community manager at Guardian Online described the community team as a three-tiered system: managing social media accounts around breaking news and exploring how to share content quickly and effectively; looking at the week ahead and trends that are emerging in terms of traffic, referrals and comments; and longer-term planning to build up "communities of interest and expertise" (Interviewee 71).

On its website, the *Guardian* described its Witness app as "the home for user-generated content on the *Guardian*". It aims to gather mobile phone footage, stills and stories, and offers the promise that the best of those submitted will be featured on the *Guardian* site, though without a fee being paid. Contributions fall into three categories: assignments (where editors put out calls on particular subjects); live news tie-ins (where contributions are sought for the liveblog); and 'open suggestions' (where the readers have the chance to suggest subjects the *Guardian* should be covering). One community co-ordinator for GuardianWitness described the app in fairly lyrical terms as part of the *Guardian's* commitment to open journalism and as a partnership.

> It's a kind of ethical thing in a way. [We] try to get audiences, get readers . . . involved in journalism, because journalists don't always know best, and people, readers, will know more things, they will see things around them, and they'll be able to improve our journalism for the benefit of all humanity.
>
> (Interviewee 77)

The app did, however, come in for initial criticism. Users had to register with the website in order to upload content and, while retaining the copyright, they had to authorise the *Guardian* to use it on the Witness site, the main *Guardian* site or in the newspaper without being paid for it, which sparked a large readers' discussion. Other media commentators compared the *Guardian's* conditions to those of Instagram, which the *Guardian* had previously criticised (Barker, 2013). Unions claimed the citizen-journalism approach was a threat to freelance journalists (Turvill, 2013).

Questions were also raised as to why those who wanted to share pictures or information would turn to GuardianWitness rather than Facebook, Twitter or Instagram to do so. There were also concerns that it did not take full account of the 'digital divide'—the fact the most privileged have the most access (and, potentially, voice). The social and communities editor acknowledged that this was a problem.

> The digital divide as well, that is always going to be something that we have to be cognisant of, and to try and compensate for. I think to not take advantage of digital communication platforms because there are some people that don't use them would also be wrong thinking.
>
> (Interviewee 72)

Yet, despite the criticisms, the *Guardian* reported that 28,000 people, representing every continent (including Antarctica), had shared a total of 60,000 photos, stories and videos within the first year alone (*Guardian* staff, 2014). While many of the photographs were not newsworthy, there were some genuine scoops. A graphic picture of the murder of Lee Rigby in Woolwich by a contributor who used the pseudonym 'Vince Lee' was used as the main photo on the *Guardian* website's live coverage (Owen and Urquhart, 2013).

The app won the innovation of the year award at the 2013 British Journalism Awards, three Lovie awards—which cover websites in the EU (GNM Press Office, 2013)—and was commended at the IPA Media Owner Awards (Faull, 2013).

The app's early success could be put down in part to a very strong community spirit among *Guardian* website readers and a tradition of collaboration and sharing content. There had been 'talkboards' at the *Guardian* as early as 1999 (Thurman, 2008), where readers shared their thoughts on news stories, as well as 'below-the-line comments' directly underneath news stories. The *Guardian*, said the community co-ordinator, had also introduced Flickr groups on the website, which had enabled readers to share photographs. GuardianWitness was one way of centralising (and, one could argue, controlling) this process. As she pointed out, there was a departure from readers as offering critiques of the *Guardian's* journalism to something different:

> It's not that the journalist does all the work and the users tell them what they've done wrong, it's more of a collaborative process from the start. At the moment, you'll see there's a lot of things to do with kittens and dogs, which is not news, so it's not just about collaborating on news stories, it is about building a community, a sense of belonging. But also it's about getting people to use the app for things that they enjoy doing, so that when they can contribute to our journalism, they've built a relationship with us. So the kittens are there, not as some cynical kitten-collection programme, but in order that people can have a practice with the app ready for when they can be involved in some news story.
>
> (Interviewee 77)

The social and communities editor was also wary about the idea of counting contributors to GuardianWitness to decide whether it had been a success or not. Her view was that Witness was shaping 'two-way' conversations between readers and journalists, and they were striving for quality not quantity.

> I would say that it's not actually about the number of people on a platform, it's about who's on the platform, and how they're making a difference. So, for example, we've had some witness assignments that really haven't had that many people submit to them, but what they have submitted is journalistically very valuable. Whereas we've had a thousand people send us pictures of buildings, and that's lovely, but it's probably not ever going to win a Pulitzer.
>
> (Interviewee 72)

The core team for Witness comprised two community co-ordinators and one assistant to review the content. The community co-ordinators would arrive at GuardianWitness between 9am and 10am to check what had been submitted overnight and review content—discarding porn, trolling posts and other material that did not fit the *Guardian's* terms and conditions. The time it took for content to be approved for use could vary from a few minutes to several hours, depending on the problems of verification or legality encountered. As the day went on, the co-ordinators reviewed more content and discussed what new assignments [requests for content] could be put up; meetings would be set up with different parts of the organisation—for example, the news desk or the sports desk—ensuring that assignments did not clash between departments or that different editors were not coming up with similar ideas.

At the beginning, GuardianWitness was such a novelty that the community co-ordinators had to ensure that desks focused on assignments that had true value.

> People across the company are really excited with what GuardianWitness can do and they want to try it out in the same way as you would want to try it out—try uploading nice pictures of buildings, whatever—but what we need is to focus their minds on the big campaigning stories, and the crowdsourcing stories that they're useful for. . . . We want to make sure it's a proper story. So that's really important to us.
>
> (Interviewee 77, community co-ordinator)

The *Guardian* has a long tradition of covering humanitarian issues. Their Katine[9] project explored development in a Ugandan village and, following on from that in 2010, their Global Development site worked closely with GuardianWitness to promote humanitarian stories and:

> provide a new space for discussion and interaction on the biggest challenges affecting the lives of billions of people across the developing world, including poverty, hunger, infant mortality, adaptation to climate change and economic development.
>
> (GNM press office, 2010)

The Guardian Development site was part of Guardian Online, but funded by the Bill & Melinda Gates Foundation, a fact that caused controversy about the blurring of lines between reporting and advocacy (Fortner, 2010; Doughton and Heim, 2011; Paulson, 2013). This controversy was fuelled when, two weeks after the launch of the website, Bill and Melinda Gates stated that they wanted to see more positive reporting of global development (Boseley, 2010).

Interviewee 78, the community co-ordinator in the Global Development section, said that the *Guardian* did face challenges and criticisms as a result of this blurring, although Interviewee 76, the deputy editor of the Development site, did not think so. She saw GuardianWitness as a natural progression, building on the *Guardian* concept of open journalism and the need to tell stories with different sources:

> It's how we drill down, report on development in a different way, try to get underneath it a little bit more. Obviously the *Guardian* has always been interested in global development, and so it was one of those natural evolutions, almost, and I think we've always wanted to get more

> southern voices, that's always been our dream to do, it's been probably the most problematic part of the site.
>
> (Interviewee 76)

Before the launch of GuardianWitness, she said that the majority of the development-related material they received was driven by NGOs. So the level of "nuanced discussion" and "authentic voice" very much depended on which aid agency they were working with. Even with very 'media savvy' NGOs this often didn't work. For example, in the Democratic Republic of Congo, the *Guardian* set up a Google Hangout facilitated by Oxfam but on top of technical difficulties, the 'real' people on the ground were uninterested in taking part. As the deputy editor of the development site put it: "They had a mountain of other things to be thinking about".

The most successful articles tended to be those by local journalists that the *Guardian* had a relationship with, such as Richard M Kavuma, editor of the *Observer* newspaper in Kampala, Uganda, or trusted NGOs.

There was also some concern that the increasing interconnectivity of the web meant some NGOs or local reporters did not fully realise the repercussions of their actions. A piece put up on the Katine website, detailing one family's expenditure for the week and picturing them outside their hut, resulted in abusive comments challenging the family's account of their spending. The *Guardian* eventually took down the piece when it became clear that these messages were not only from strangers, but also from other villagers who had seen the piece on village computers. The deputy editor of Global Development said:

> There have been things quite recently that we haven't used actually from a reporter, because I wasn't convinced that they had got permission from the NGO, or the NGO's said that they have, but I just wasn't convinced that that person who had been raped would be happy to have their name shared. With the web, it's not just Katine, it's around the world, and do the people you're interviewing fully get that this is going to be seen by millions of people—including the person that they live near who might happen to go and look at the computer. So it's trying to have all these checks and balances, which is a challenge and not straightforward at all.
>
> (Interviewee 76)

The Global Development site had also experimented with Flickr photograph groups, usually with straightforward asks such as calls for messages for International Women's Day. *Guardian* research, however, suggested that the readership remained primarily amongst NGOs and academics, rather than the ordinary people they hoped to reach:

> The studies, the research that we've done, I think we have got more general public in now, but the majority are still people who are working

> in the industry to some degree, whether it's studying or academia or NGOs or other development journalists.
>
> (Interviewee 76, deputy editor Global Development)

Meanwhile, Interviewee 75—who was in charge of GuardianWitness—saw one role of the app as challenging the typical coverage of humanitarian crises. The pictures they received were not usually used in the paper itself, but in the liveblog. She introduced a three-month window for Syrian refugees to send in pictures for one particular liveblog, in the hope of maximising the number who did so:

> This we feel is a good space just get to know about various sides of what's going on, what's it like in a refugee camp. Someone sent in a picture about inadequate water waste in one of the transit camps in northern Syria. We get really nice things like girls laughing, that's really important to see that as well, because you just don't want to see this traditional line of refugees, hopeless. It's good to see these people as real people, I think.
>
> (Interviewee 75, community co-ordinator GuardianWitness)

Examples included pictures of Syrian refugees living in derelict buildings in Lebanon[10]; aid agencies teaching refugees about sexual and reproductive health; a small boy who had set up his own shop; and young men setting up a barber shop. This challenged the 'compassion fatigue' imagery of long queues at feeding points, or refugees sitting inside a tent with nothing to do.

> We know lots of these refugees are living inside cities, and Lebanon has no camps, so this is about a five-storey house. . . . Every city in Lebanon is absolutely crammed with refugees, and this of course is a very nice image showing that, and it wouldn't make the front page on its own, but it provides another aspect of the story when it's here.
>
> (Interviewee 75)

Thus GuardianWitness offered an opportunity to confront the usual stereotypes of refugee life. The picture of the young men in the barber shop was seen as significant, for example. It reflected the fact that grooming was still important, even in a world where people had lost most of their possessions.

Yet the aim of having mainly stories from the refugees themselves did not work out as planned—the vast majority of 'eyewitness' photographs came from NGOs or professional photographers. Partly, this was because the app was so new:

> Ideally you'd go in with a journalist, and say we've got this thing [that is, the GuardianWitness app] so you can use it. This is brand-new, though, we're just getting everyone used to the idea we have this tool,

> and helping people realise that they can continue sending us their stories, even when the journalist has left, which is absolutely paramount to me. Because the story isn't about when the journalist arrives and leaves, it's about what carries on after, and what's happening before. That's why this is so cool.
>
> (Interviewee 75)

Those framing the stories remained primarily NGOs, however. First, because GuardianWitness in its earliest stages, when still not that well known, had reached out to NGOs in order to gather material. Second, because the requests for pictures were primarily via Twitter and Facebook, which the NGOs followed. Third, because of a new savviness Interviewee 75 had seen among NGOs, who realised that the old-fashioned branded T-shirts and warehouse shots would not work in this context.

> They're the ones who sent this stuff in, and I think actually they have gone away from "look at us giving all our stuff out". . . . And this is what the whole point of this is. So these guys, they're smart, so I think they do understand what we're trying to do.

This was not the first time that the *Guardian* had run into difficulties trying to use local voices. It had faced similar problems with Katine. While the three-year-old project was seen as innovative, various issues emerged—not only about the media organisation's dual role as reporter and solicitor of money, but also about the kind of user-generated content the paper received when encouraging the villagers to photograph items or events important to them. As Sarah Boseley of the *Guardian* reflected, when it gave out disposable cameras to the villagers in the hope of getting a new perspective, "most of them", she said, "just took pictures of their cows" (Cooper, 2009, 2011).

It can be argued, of course, that these were the most important items to villagers scraping a living from farming, and "cattle wealth" has a particular resonance (Jones, 2009). Boseley's remark, however, indicated the difficulty in reconciling the Western media's view of what would make interesting, newsworthy images, and what Katine's residents saw as most important.

Six years on, however, the GuardianWitness view was that there had been a tremendous growth in Western media literacy among people facing poverty or humanitarian disasters, helped by the spread of social media. There was much more understanding about what kind of pictures would work well on a website and, crucially, the importance of being able to verify these images.

> There is definitely a media literacy. With Syria, they're aware that verification is a massive thing for them, so for all sides, you'll find when they're sending us something, "oh, this bomb was dropped by

> government forces", they'll hold up a newspaper.[11] I think the Arab Spring has changed a lot of stuff because they know we have to verify.
>
> (Interviewee 75)

Like many media organisations, there was at that time a rather casual attitude towards the verification of UGC, with different staff taking different approaches (Cooper, 2013). Interviewee 75 said there had been few problems around verification for GuardianWitness because, in the main, she had been able to contact the producers of content by phone. However, she saw this as very much as only a beginning: "That level [is the] first level of verification, you just don't know what will come in in the future, do we?"

Interviewee 77 gave more detail on the *Guardian*'s approach to verification, saying it had employed Claire Wardle, a social media specialist, to come in and advise on a 'two-step approach'. For subjects that seemed non-controversial "such as a kitten or a building", she said, "we'll generally give it a quick look, make sure there's nothing in it that shouldn't be, give the username a quick look and read the description. If it seems fine then we publish it". For more sensitive visuals, the process was longer. The co-ordinators would do a reverse image search on sites like TinEye or Google Images and check out the creator's email address. They would watch videos all the way through to check for copyright issues, and with stories like Syria, they would attempt to look at satellite images online, to locate the exact place the video was taken.

In these more controversial cases, the 'gold standard' would be to speak to someone by phone. The GuardianWitness journalists would also try to seek verification from staff at the *Guardian* itself.

> We're calling it two-step verification because . . . the second step will be to give it to the desk, to say, look, we know this, we know this, we know this, but you're the subject specialist, you know that dialect of Arabic, you know what people were wearing, you'll be able to tell us if it looks right or not.
>
> (Interviewee 77)

In other areas, such as security, there had been few issues in the beginning, thanks to greater understanding by NGOs of how media organisations used UGC. But Interviewee 75 said a lot of the content could not be used because "it was too dangerous" (for those identified in the photographs)—by the time the first 50 pictures had been put up on the site when it launched, around the same amount had been rejected. The social and communities editor (Interviewee 72) was also less gung-ho, and saw a potential for ethical problems to emerge because of the kind of content GuardianWitness might encourage:

> Are we putting people into danger by covering an assignment? We need to think. When we think people might consider putting themselves into

> a dangerous situation, we ask them not to. But ultimately one day a picture might come in on something, say on refugees, and it's distressing, and it's quite evidently from somewhere dangerous, but it has a news value. What do we do about that? So there's ethical concerns raised about it.

The community co-ordinator in Global Development highlighted other ethical problems with asking people to upload pictures to an app like this. She felt that many of her contacts still preferred to deal primarily in a more personal way, such as via email. For those involved in stories such as Syria or other humanitarian disasters, it could seem quite a cold way to share content:

> Quite often there's things with photos and especially if they're emotional stories and actually sometimes people do need a little bit of reassurance about "yes, this is great to use; no, we're not going to use that". Actually, I think people have a fear of rejection as well, and, just because of the way GuardianWitness is set up, there's no quick way to tell people why we can't use their photos. So if we can have that link with people on email then it makes it a bit easier if they have any questions.
>
> (Interviewee 78)

The *Guardian* had laid down several rules for the new app. The most important of these was that they would not use pictures of under 16s on the site because of privacy concerns. This was admirable, but it proved particularly problematic for humanitarian stories.

> I've got about 35 wonderful images of children that I can't use right now, but actually as far as the aid agencies are concerned, it's safer sending in images of children than it is of the adults.[12] It's not ideal for them, and we're trying to change that, actually. Everyone's getting round it.
>
> (Interviewee 75)

There were also particular cultural issues—where images depicted grief and even death—that GuardianWitness, as a global website based in London, struggled with. While there has been growing concern about the rise in visibility of death in the news, such images are in fact extremely rare (Zelizer, 2010; Hanusch, 2013; Morse, 2014); 'death photos' tend to be implied/presumed, fractional or masked.

> If I got this amazing image, but I thought it was inappropriate . . . we wouldn't show in this country grieving in the way we would in other countries, for instance, and I am quite aware there are some things you've just got to be sensitive about, think of the impact on people it's happened to.
>
> (Interviewee 75)

For GuardianWitness, the problem was balancing Western ideas (and within that, in Hanusch's terms, northern European/American ideas with Latin American/southern European) and non-Western ideas about what was acceptable.

Conclusion

In an age of smartphones, mediated witnessing has become the norm. But journalists seek to 'normalise' (Singer, 2005) this challenge to their journalistic authority in very different ways, as this chapter has shown. Boundary work has traditionally looked at how stakeholders aim to impose a dominant definition of what it is to be a journalist, or perform acts of journalism, in order to be seen as legitimate. Drawing a line between insiders and outsiders is clearly key.

However, what is interesting is how the permeability of journalism and its relative weakness as an autonomous field (Bourdieu, 2005) has led journalists to seek different solutions. First, many media organisations chose to co-opt and clone the tropes that characterised UGC to benefit from its attractiveness to the audience. By adopting content creators' methods of communication (Twitter, mobile phone pictures, blogs, Vines), they attempted to appeal to an audience who might have otherwise turned to other mediators.

For the *Guardian* and the BBC, however, there was a more organised approach to dealing with the challenges of UGC.

The BBC clearly had more experience and resources compared with GuardianWitness, but in many ways their aims were similar: to ensure that the best and most timely user-generated content was relayed to the general public via the organisation.

However, the BBC's focus was in some ways more traditional; most UGC was seen as the basis for case studies. While text could be used for liveblogs as well as live programming, it was seen as a starting point—the ultimate aim being to get an exclusive BBC interview. As Interviewee 88, a senior broadcast journalist at the BBC describes, for example, during the Haiti earthquake, the value of Carl Pedre's tweets was acknowledged, but what the BBC journalists were really aiming for were Skype or other forms of live interview with him in the field. For them, UGC was primarily another source to widen their news coverage, but in a way that upheld the traditional primacy of the journalist as gatekeeper and interpreter. While video, stills and tweets were used on live programming as well as online, it was frequently as an introduction to an interview.

Some of this was undoubtedly due to the BBC's commitment to verification. In terms of how this altered the power dynamics in the field, however, the corporation was still very much in control. The creators of UGC were absorbed into different news programmes. As the day wore on, they were increasingly likely to be used in vox pop interviews with BBC journalists, rather than the BBC linking to their actual content, unless it was particularly dramatic.

The *Guardian* had devoted considerable time and resources to encouraging more user-generated content via its new app, which proved popular with *Guardian* readers until its closure at the end of September 2018.[13] This worked well with straightforward and light-hearted content but proved more difficult in complex stories such as humanitarian crises.

There appeared to be a commitment, however, to find different voices and to tell stories that went beyond the traditional template reporting complained of by Alagiah (1998). Who then had control of the stories and the pictures was something that the staff found difficult to explain. The criticisms levelled at GuardianWitness before launch clearly still rankled with staff, but it remained the case that the site was effectively getting content for free, which it then branded exclusive to the *Guardian*.

The journalists' defence was that they were encouraging people in vulnerable situations to tell their stories to a wider audience. Yet, problems remained in reaching out to those at the centre of these stories. Clearly, the *Guardian*, as an Anglophone publication, missed some key voices even in disasters in the developed world. As Interviewee 71 put it:

> We've tried some experiments in terms of translating our content, and our callouts, but it's minimal, and so getting the message out to the right people can be really difficult, and equally that then goes against the thing which I mentioned earlier which is that you want to find real and alternative voices to our coverage. If you can only do that and you're reaching out to an expat community, not so good.

As a result, in terms of field theory, the new agents gaining access to the field in this case were less likely to be those who had experienced a crisis themselves, and more likely to be NGOs with their own particular agendas to express. This was something that GuardianWitness accepted and co-operated with in order to get the content they needed.

However, while this was the norm, both the Guardian and the BBC did use tweets, images, YouTube, Audioboo at times of acute humanitarian disaster however. As observed in this chapter, many of those pieces of content were from aid agencies and other authority figures but they did also encompass ordinary survivors. The next chapter looks at who they were, and crucially how those creators found the use of their words and texts.

Notes

1. Named after CNN correspondent Dr Sanjay Gupta who performed a dual role during his time in Haiti following the 2010 earthquake. Gupta, a trained doctor, not only reported on the quake but was filmed carrying out medical procedures and first-person pieces to camera (see Farhi, 2010; Chen, 2010).
2. Interviewees 36, 38 and 40 all have 'editor' in their titles but this indicates seniority in the newsroom; their jobs were as correspondents.
3. Tempero (tempero.co.uk) is a social media management firm which was employed by the BBC to moderate comments on the BBC website.

4. A case study in journalistic parlance refers to an in-depth interview with a member of the public to illustrate a particular story or theme.
5. Interviewee 87 specifically named Tamil groups in Sri Lanka and "liberal English-speaking types" in Turkey during the time of the 2013 Gezi Park protests.
6. On-screen captions (named after the Aston Broadcast System).
7. A reverse image search is technology used by a search engine, which takes an image file as input query and returns results related to the image, so you can see where else the image appears on the web.
8. A Q&A with former Guardian editor Alan Rusbridger where he discusses the Guardian approach can be found here: www.theguardian.com/commentisfree/2012/mar/25/alan-rusbridger-open-journalism.
9. Katine was a three-year project in a village in Uganda when the paper linked up with two NGOs—the African Medical and Research Foundation (Amref) and Farm Africa—to report on their progress and encourage donations to the work, which was match-funded by Barclays.
10. The photograph by Melissa Fleming of UNHCR features on the 25 July liveblog at 12.59pm.
11. She is referring to the practice of holding up the day's newspaper to prove that an event happened on the day claimed.
12. Interviewee 75 was referring to the fact that a) children were often less identifiable and b) were less likely to have been photographed in a controversial way.
13. Guardian readers were told that after that date they would be able to contribute UGC via the Community Section.

References

Abbott, A. 2014, *The system of professions: An essay on the division of expert labor*, University of Chicago Press, Chicago.

Alagiah, G. 1998, 27–28 May, *Dispatches from disaster zones: Reporting humanitarian disasters*, Conference, British Red Cross, London.

Allan, S. 2006, *Online news: Journalism and the internet*, Open University Press, Maidenhead.

Allan, S. 2013, *Citizen witnessing: Revisioning journalism in times of crisis*, Polity Press, Cambridge.

Allan, S. & Thorsen, E. (eds). 2009, *Citizen journalism: Global perspectives*, Peter Lang, New York.

Barker, D. 2013, 7 April-last update, *The Guardian's terms & conditions: Worse than Instagram?* [Homepage of Dan Barker], [Online]. Available: http://barker.co.uk/guardiancopyright [2015, 9 June].

Belair-Gagnon, V. 2015, *Social media at BBC news: The re-making of crisis reporting*, Routledge, New York.

Bell, M. 1998, "The journalism of attachment", in *Media ethics*, ed. M. Kieran, Routledge, London, pp. 15–22.

Boaden, H. 2008, 13 November-last update, *The role of citizen journalism in modern democracy* [Homepage of BBC], [Online]. Available: www.bbc.co.uk/blogs/theeditors/2008/11/the_role_of_citizen_journalism.html [2015, 16 September].

Boseley, S. 2010, 27 September-last update, *Bill and Melinda look on the bright side* [Homepage of The Guardian], [Online]. Available: www.theguardian.com/society/sarah-boseley-global-health/2010/sep/27/maternal-health-bill-and-melinda-gates-foundation [2015, 10 June].

Bourdieu, P. 1998, *On television and journalism*, Pluto, London.

Bourdieu, P. 2005, "The political field, the social science field, and the journalistic field", in *Bourdieu and the journalistic field*, eds. R. Benson & E. Neveu, Polity Press, Cambridge, pp. 29–48.

Brandtzaeg, P.B. & Chaparro Domínguez, M.Á. 2018, "A gap in networked publics? A comparison of younger and older journalists' newsgathering practices on social media", *Nordicom Review*, vol. 39, no. 1, pp. 1–15.

Cooper, G. 2009, 30 January-last update, *Katine offers new ways to campaign* [Homepage of The Guardian], [Online]. Available: www.theguardian.com/society/katineblog/2009/jan/30/glenda-cooper-new-media [2015, 9 June].

Cooper, G. 2011, *From their own correspondent? New media and the changes in disaster coverage: Lessons to be learned*, Reuters Institute for the Study of Journalism, Oxford.

Cooper, G. 2013, 25–29 June, *Heading for a disaster? Ethical and legal questions raised when mainstream media use user-generated content to report humanitarian crises*, Conference paper IAMCR, Dublin.

Doughton, S. & Heim, K. 2011, 23 February-last update *Does Gates funding of the media taint objectivity?* [Homepage of Seattle Times], [Online]. Available: www.seattletimes.com/seattle-news/does-gates-funding-of-media-taint-objectivity [2018, 27 August].

Dubberley, S., Griffin, E., & Bal, H.M., 2015 *Making secondary trauma a primary issue: a study of eyewitness media and vicarious trauma on the digital frontline,* Eyewitness Media Hub, Istanbul. Available: firstdraftnews.org/wp-content/uploads/2018/03/trauma_report.pdf [2018, 27 August].

Farhi, P., 2010, 20 January-last update *In Haiti, reporters who double as doctors face a new balancing act* [Homepage of The Washington Post], [Online]. Available: www.washingtonpost.com/wp-dyn/content/article/2010/01/19/AR2010011904293.html [2018, 27 August].

Faull, J. 2013, 19 November-last update, *Yahoo, ITV.com, The Guardian and Sky Media among the winners at the IPA 2013 media owner awards*. Available: www.thedrum.com/news/2013/11/19/yahoo-itvcom-guardian-and-sky-media-among-winners-ipa-2013-media-owner-awards [2015, 9 June].

Fortner, R. 2010, 8 October-last update, *The web grows wider* [Homepage of Columbia Journalism Review], [Online]. Available: www.cjr.org/the_observatory/the_web_grows_wider.php [2015, 10 June].

Gieryn, T.F. 1983, "Boundary-work and the demarcation of science from non-science: Strains and interests in professional ideologies of scientists", *American Sociological Review*, vol. 48, no. 6, pp. 781–795.

Gieryn, T.F. 1999, *Cultural boundaries of science: Credibility on the line*, University of Chicago Press, Chicago and London.

GNM Press Office. 2010, 14 September-last update, *The Guardian launches global development website with Gates foundation* [Homepage of The Guardian], [Online]. Available: www.theguardian.com/gnm-press-office/guardian-launches-global-development-site [2015, 10 June].

GNM Press Office. 2013, 4 November-last update, *Lovie awards 2013* [Homepage of Guardian Online], [Online]. Available: www.theguardian.com/gnm-press-office/guardian-witness-wins-at-lovie-awards-2013 [2018, 5 April].

Guardian Staff. 2014, 27 March-last update, *Guardian witness awards winners announced* [Homepage of The Guardian], [Online]. Available: www.theguardian.com/media/2014/mar/27/guardian-witness-awards-winners-announced [2015, 9 June].

Hanusch, F. 2013, "Sensationalizing death? Graphic disaster images in the tabloid and broadsheet press", *European Journal of Communication*, vol. 28, no. 5, pp. 497–513.

Harrison, J. 2010, "User-generated content and gatekeeping at the BBC hub", *Journalism studies*, vol. 11, no. 2, pp. 243–256.

Jones, B. 2009, 15 September-last update, *Cattle: A particular sort of wealth in Katine* [Homepage of The Guardian], [Online]. Available: www.theguardian.com/society/katineblog/2009/sep/15/livelihoods-money [2015, 9 June].

Kitch, C. 2008, "Placing journalism inside memory—and memory studies", *Memory Studies*, vol. 1, no. 3, pp. 311–320.

Lamont, M. & Molnár, V. 2002, "The study of boundaries in the social sciences", *Annual Review of Sociology*, vol. 28, no. 1, pp. 167–195.

Matheson, D. 2004, "Weblogs and the epistemology of the news: Some trends in online journalism", *New Media & Society*, vol. 6, no. 4, pp. 443–468.

Morse, T. 2014, "Death images in Israeli newspapers: Ethics and praxis", *Journalism Studies*, vol. 15, no. 1, pp. 99–113.

Owen, P. & Urquhart, C. 2013, 25 April-last update, *Woolwich: New video shows suspect charging at police* [Homepage of The Guardian], [Online]. Available: www.theguardian.com/uk/blog/2013/may/24/woolwich-new-video-shows-suspect-charging-at-police-live-updates [2015, 9 June].

Paulson, T. 2013, 14 February-last update, *Behind the scenes with the Gates foundation's 'strategic media partners'* [Homepage of Humanosphere], [Online]. Available: www.humanosphere.org/basics/2013/02/a-personal-view-behind-the-scenes-with-the-gates-foundations-media-partners/ [2015, 10 June].

Seaton, J. 2005, *Carnage and the media: The making and breaking of news about violence*, Allen Lane, London.

Singer, J.B. 2005, "The political j-blogger 'normalizing'a new media form to fit old norms and practices", *Journalism*, vol. 6, no. 2, pp. 173–198.

Star, S.L. 2010, "This is not a boundary object: Reflections on the origin of a concept", *Science, Technology & Human Values*, vol. 35, no. 5, pp. 601–617.

Star, S.L. & Griesemer, J.R. 1989, "Institutional ecology, translations' and boundary objects: Amateurs and professionals in Berkeley's museum of vertebrate zoology, 1907–39", *Social Studies of Science*, vol. 19, no. 3, pp. 387–420.

Thurman, N. 2008, "Forums for citizen journalists? Adoption of user-generated content initiatives by online news media", *New Media and Society*, vol. 10, no. 1, pp. 139–157.

Turvill, W. 2013, 24 April-last update, *Unions claim Guardian witness project threatens freelance journalists* [Homepage of Press Gazette], [Online]. Available: www.pressgazette.co.uk/content/unions-claim-guardian-witness-project-threat-freelance-journalists [2015, 9 June].

Wahl-Jorgensen, K. 2015, "Resisting epistemologies of user-generated content? Cooptation, segregation and the boundaries of journalism", in *Boundaries of journalism: Professionalism, practices and participation*, eds. M. Carlson & S.C. Lewis, Routledge, New York.

Wardle, C. & Williams, A. 2008, "ugc@BBC:Understanding its impact on contributors, non-contributors and BBC news", *BBC*, Available: www.bbc.co.uk/blogs/knowledgeexchange/cardiffone.pdf [2018, 28 June].

Zelizer, B. 2010, *About to die*, Oxford University Press, Oxford.

7 'Tweeting the Quake'

How Ordinary Citizens Tell Their Stories

> [Reproducing my] tweet ended up boosting my followers by at least 100 in an hour. [. . .] I realised I'd better stop joking and felt a responsibility to tweet actual news.
>
> (Interviewee 48, music writer)

> I was pretty stressed about what I had seen [the Oklahoma hurricane] and felt guilty about my part in taking videos of it. I did not profit in any way from this and I have found it awkward in wanting to go back to using social media.
>
> (Interviewee 65, security guard)

The two quotations above are from people who could be described as 'accidental journalists' (Allan, 2013)—eyewitnesses who created user-generated content in two recent humanitarian crises. The first tweeted during the 2011 Great East Japan earthquake; the second took a Vine video of the 2013 Oklahoma hurricane. Both posted content on the internet with little realisation that their pictures or words would go beyond a small circle of friends; in fact, they both went global.

This chapter deals with voices that are not usually heard in discussions about user-generated content and the media: those whose content is taken and appropriated by the mainstream media. Most research into user-generated content (UGC) has concentrated on the impact that it has had on journalists and what they do (Hermida and Thurman, 2008; Paulussen and Ugille, 2008; Singer and Ashman, 2009; Singer, 2010; Harrison, 2010). The reaction of audiences to interactivity and UGC has also been well-documented and theorised (Hujanen and Pietikäinen, 2004; Bergstrom, 2008; Wahl-Jorgensen, Williams and Wardle, 2010; Larsson, 2011). What creators of content—dubbed 'the people formerly known as the audience' (Rosen, 2006)—feel about the use of their work by the mainstream media is less well known. Some research has suggested that UGC creators may wish to establish their own norms and values (Robinson, 2010), and that while they are willing to take part in the creation of content, the continued gatekeeping role of journalists (Bruns, 2008) once the content has been handed over leads to a sense of dissatisfaction among UGC creators (Borger, van Hoof and Sanders, 2016).

This research, however, has tended to focus on 'below-the-line commentators' rather than those who find their work replicated in the mainstream media, and who are the subjects of this study. The UGC creators interviewed were asked about whether permission was sought or payment offered (or not offered, in many cases), and about privacy issues.

The speed at which newsgathering has absorbed and utilised content created by non-journalists has been surprising. One journalist admitted hearing people once say: "'Why would we want to use this? Look at the quality of mobile footage; who would be interested in it?' Now when a major story happens everyone is beating at the UGC door" (Wardle, Dubberley and Brown, 2014:5).

While 'citizen journalists' (those who habitually create content with a journalistic approach) or non-professionals (activists, aid agencies, etc) may have tweeted, filmed or taken photos of events, much of this content is shot by people dubbed 'accidental journalists'—those who simply happened to find themselves caught up in a story and were "in the right (or wrong) place at the right (or wrong) time" (Wardle, Dubberley and Brown, 2014:5).

This chapter is based on quantitative analysis of liveblogs, which used UGC, and qualitative interviews with those who found themselves in the traditional role of the witness once associated primarily with journalists. It is divided into two parts. First, it looks at how prevalent the use of UGC is during catastrophic events and what types of producers find their content used. It questions whether framing devices that appear to privilege UGC are in fact using content created by those outside the mainstream media or other powerfully privileged users such as NGOs, celebrities, government and quasi-governmental organisations. In a crisis such as the Great East Japan earthquake, for example, the study examines how many 'accidental' journalists' voices were actually heard, and to what extent they were absorbed and superseded by traditional voices that may have adopted new media technologies to preserve their domination and gatekeeping role.

Second, this chapter examines what the consequences were for those eyewitness publishers in terms of ethical questions of permission and privacy; and how democratising or (dis)empowering this potentially was for them. This is what I have dubbed 'the virtual doorstep': the experience that many of these UGC creators had of journalists putting their content (and lives) online. The UGC creators featured were not primarily putting content up on social media sites or sending it to news sites to make money. Rather, they were sharing what they had seen with friends and family via the social web in the shock of the aftermath of a humanitarian disaster.

Even those who were journalists—one, for example, was a music writer—were not writing or filming for their own organisation. They were sharing information online out of a sense of public service or—with the human need to share—simply the feeling that it was important to do so.

To examine more closely the effects of this, as well as looking at the blogs, it also includes three further case studies whose content had gone viral: Interviewee 65, whose Vine videos had been a crucial part of the 2013 Oklahoma hurricane coverage, and Interviewee 66 and 67, whose pictures and videos of the 2013 Vauxhall helicopter crash had sparked controversy. I wanted to compare and contrast the experience of the Japan UGC creators with that of those who had had a much more intense reaction. The journalists at the BBC and the Guardian, who recommended these stories, saw them as game-changing moments—particularly the Vauxhall crash—because of the quality of the pictures and because the increasingly rapid news cycle meant that they chose to use the pictures and then seek permission. They also saw them as examples where the ethical questions I was asking were apposite:

> There's an interesting case where the helicopter crashed into the skyscraper on the Thames a few months ago. The best pictures were pictures that people were tweeting. People were taking, I think, some really professional-standard pictures. So we had a choice there, which was basically use it, and then deal with the consequences afterwards, or don't run with it. I particularly felt that we should use it.
>
> (Interviewee 70, liveblogs editor, *The Guardian*)

> It's a nightmare situation to be in [the Oklahoma hurricane]. The journalist in me says you need that story, you need those people who were there. But the person in me says I wish they just had some time to get around what on earth is going on, and there wasn't, it was just 'break it, break it, break it', you know?
>
> (Interviewee 86, UGC producer, BBC)

The Liveblog and 'Primary Twifiner'

To try to establish the kind of content—and the kind of content producers—that appear during a rapid-onset humanitarian disaster, I chose to focus on an increasingly popular mainstream media device: the liveblog. The liveblog is a relatively new web-native format, which is now being used by newspapers on their websites and by broadcasters, such as the BBC, to produce a synthesis of traditional journalism and contemporary digital technologies (Thurman and Walters, 2013).

In the UK, the liveblog can be traced back to the *Guardian* newspaper in 1999, although for a decade it was used mainly for sports rather than breaking news. 'News' liveblogs started to emerge in the mid-2000s. Kevin Anderson's 2004 US election blog incorporated readers' comments and the *Guardian* ran a liveblog covering the 7/7 terrorist attacks in London in

July 2005 (although the first time readers' comments were integrated to a *Guardian* blog was in 2007 (O'Mahony, 2014)).

Today, the ubiquity of the liveblog spans from left-wing tabloids like the *Mirror* to right-wing broadsheets like the *Daily Telegraph*. It has been used to cover subjects as diverse as the Arab Spring, the Boston marathon bombings, Prime Minister's Questions and even the announcement of the Duchess of Cambridge's second pregnancy (Tran, 2014). Liveblogs are often much more popular with the public than traditional news stories and picture galleries, and people tend to spend longer on a liveblog (Thurman and Walters, 2013).

But, while the liveblog—with its quick-fire updates and reliance on huge amounts of content for breaking news stories—might seem an ideal place for creators of UGC to have their voices heard, does the content that ends up on the blog reflect that? A writer of one of the BBC's liveblogs (called live pages) attempted to explain what the blogs should consist of:

> A good live page should be engaging, informative, fast moving, comprehensive and clickable (i.e. have lots of URL links). So I always aim to have a good mix of official sources/reaction, people on the ground, ordinary punters, analysts, pictures/galleries and tweets.
>
> (Interviewee 102, BBC 'Core' writer)

There were no specific livebloggers at the *Guardian*, but general reporters were made responsible for the blog on a shift basis. They were given training but often did not enjoy the role of liveblogger—mainly because it meant they were stuck in the office, rather than out in the field, and because they could not take a break or pursue their own stories. For them the focus was on getting their own words in the blog, but also the idea of having unmediated reactions to events:

> Something like the Middle East live blog will rely a lot on video footage, because that's part of trying to get that story out of Syria. . . . Tweets we do use a lot. It's such an interesting and new thing—you can sort of get access to what people are saying without the filter of their press officer or spokesperson. For politics, that's brilliant.
>
> (Interviewee 70, liveblogs)

Most editors named UGC as most useful in the immediate aftermath of a story—before a journalist could get there. But the kind of content used in the liveblogs I examined often comes from traditional sources, rather than the eyewitness accounts that journalists seemed to be seeking.

For example, the *Guardian*'s liveblog of the Great East Japan earthquake starts at 7.22am GMT. But the first time we hear anyone speak who is not allied to government, media organisation or an NGO is nearly two and a half hours later at 9.41am, when a tweeter has his audioboom (a web application that allows users to record and share sound files) reproduced on the *Guardian* site.

There are very practical reasons why it may take time for UGC to emerge; for those caught up in disasters, their main priority is to get to safety. But in the meantime, we have heard from the *Guardian*'s Tokyo correspondent, Reuters, Al Jazeera, the BBC, the US Geological Survey, the UN, the Japanese prime minister (via wires), the International Federation of Red Cross and Red Crescent Societies (IFRC) and the International Tsunami Warning Centre. Also reducing reliance on UGC are 'below-the-line' commentators—those who have already registered with the *Guardian*, whose journalists can see how much they have posted in the past or whether they have been blocked for abusive behaviour. Again this prioritises the 'helpful source' but is selective; these below-the-line commentators are *Guardian* readers, rather than the general public.

Yet this is a rapidly changing medium. Two years later, at the time of the Oklahoma hurricane in 2013, the *Guardian* liveblog was already using a very different approach. The increasing focus towards mobile and away from computer users meant that there are far more pictures and footage. Below-the-line commentators were now interviewed as part of a piece rather than their original text being used. The noise on social media has become so loud that the *Guardian* liveblog editors said that they now use a stream of 'verified users' in the early stages of a breaking news story—meaning that the editor largely depends on other 'verified' users retweeting 'unverified' information. So the institutions that have embraced Twitter—for example, the Federal Emergency Management Agency (FEMA), the Red Cross and even the Queen—tend to receive more coverage as a result.

I contacted those whose content was used—knowingly or otherwise—for liveblogs at a UK broadsheet newspaper (the *Guardian*) and broadcasting organisation (the BBC) on the first day of the Japanese quake of 2011. Around 23 individuals responded to my request for semi-structured interviews via phone, Skype or email. They tended to be younger than the UGC contributors identified in other studies (Wardle and Williams, 2008), to work in digital-based industries and to be mainly white British or American (this is unsurprising because they were commenting on—or used on—a British newspaper's website and Western TV stations).

The Great East Japan earthquake Blog 2011—the Guardian

The *Guardian* blog of the Great East Japan earthquake began at 7.22am GMT on Friday 11 March with a picture of Sendai credited to the AP and an elongated entry from the *Guardian's* Tokyo correspondent Justin McCurry. It reads like a traditional news story, giving information about the earthquake striking north-east Japan and triggering warnings of tsunamis.

There are 76 timed updates throughout the day. Most of these involve information from a single source, or a news round up, although 14 do include more than one source.

Of the 76 timed updates made on this liveblog, 14 appear to be based on UGC, either wholly or partially. I attempted to contact all the UGC contributors by various means. I used Google and Twitter searches to track down those who had used similar names. I messaged the YouTube contributors via YouTube. I also asked the *Guardian* communities team to contact some of the below-the-line commentors on my behalf. As a result I was able to communicate with seven contributors in depth via email or Skype who were mainly aged in their twenties and thirties and who had jobs such as English teachers, project managers, engineers and music writers.

There were 14 updates from *Guardian* correspondents and 14 UGC sources. However, the biggest contributors to the blog, with 26 entries, were the newswires—AP and Reuters in particular—as the blog took breaking headlines and reproduced them. There were also a significant number of authority figures (24) who were quoted or alluded to. Because of the difficulty of sourcing pictures and information quickly and accurately there were 16 references to other media outlets such as the BBC and the Japanese state broadcaster, NHK. There were only two references to aid agencies—the ICRC, and Oxfam. This reflects the relative slowness of aid agencies to exploit this form of social media and the fact that in the first hours of a breaking news story of a disaster the focus tends to be on the unfolding crisis rather than the aid that is needed to alleviate it.

The UGC on the blog included a YouTube video of Sendai as the earthquake struck and video footage of Tokyo shaking. It also included a film of the tsunami, which appears to have been made for Japanese TV, some YouTube footage which no longer exists and a live webcam. There is one audioboom—from Interviewee 44, an English teacher—five below-the-line comments and three tweets.

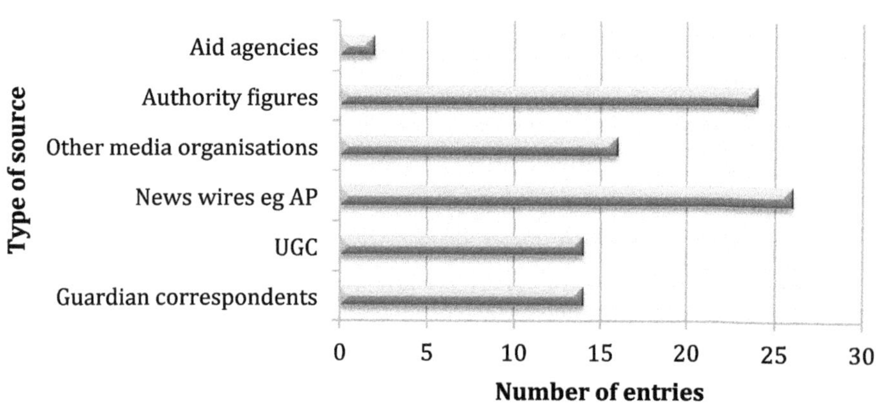

Figure 7.1 Entries on the *Guardian* Liveblog 11 March 2011

The statistics show the liveblog follows Schlesinger's theory of the 'helpful source'—often the official government/business spokesperson, who has the resources to staff press offices around the clock when breaking news happens (Schlesinger, 1978).

In covering a humanitarian disaster, sources can also include the eyewitnesses on the ground, with journalists searching for geolocated tweets. But the drive towards speeded-up news rotation also privileges 'authority' figures—a wider group than Hall et al's 'primary definers' (1978). Prime ministers and presidents still have instant access (Gans, 1979) but now they can have it via their verified Twitter accounts and journalists' inboxes.

The Great East Japan earthquake Live Page 2011—the BBC

The BBC's Great East Japan earthquake blog had 229 updates compared to the *Guardian's* 76, but it did not have significantly more UGC. The BBC's UGC is divided into two main types: tweets and information sent in by viewers to the BBC's *Have Your Say* site. The BBC has its own UGC 'hub', which actively trawls for user-generated content, but most of this content is used for programmes rather than being put on the liveblog. Apart from a livestream of waves (which also appeared on the *Guardian* site) and a picture posted on Twitter by an actor (Interviewee 62), the UGC component of the BBC blog was overwhelmingly text based. This reflects the focus on visual and audio information for programmes and the difficulty in sourcing UGC in the immediate aftermath of a disaster. With many Japanese commenting on social media in Japanese, the language barrier is also a problem.

The BBC's UGC came mainly from *Have Your Say*—the equivalent of the below-the-line comments on the *Guardian* site—which asks readers to specifically respond to a particular news story. Of the comments, 12 came via *Have Your Say*, 12 via Twitter and one was a livestream of waves, which was accessed via emptyage.tv but came originally from Twitter (the same webcam that the *Guardian* linked to).

Of the 227 posts on the liveblog, 25 could be classed as user-generated content. These were mainly generated via either *Have Your Say* on the BBC website, the UGC hub or Twitter.

The BBC had not kept records of all the people they were in touch with via *Have Your Say* and, while the UGC hub did keep records of some contacts, these included ones that had been used for different programmes, not necessarily all on the liveblogs. I contacted the names they gave me regardless, to try to build up as rich a picture as possible as to how the BBC used UGC on that day. I tried to find those who came via the *Have Your Say* website through Twitter, Google search and interviewed 16 ranging in age from their twenties to late forties, and doing jobs such as teaching, student, small business owner, tourism worker, professor and IT manager.

The BBC calls its liveblogs 'live pages', or LPs for short, and tends to use them for a developing story. They are written by so-called 'core' writers,

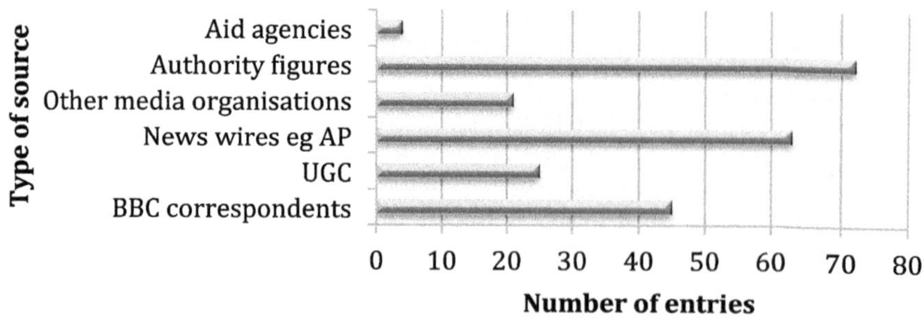

Figure 7.2 Entries on the BBC Live Page 11 March 2011

and there are several for each blog. The core writer on the Japan blog commented on their role as follows:

> Basically, the idea of a live page is a 'one-stop-shop' (news hub) where you get everything moving along fast with the story. . . . My main duty was to get the feeds (checking wires; watching/listening to BBC/non-BBC correspondents; checking other non-BBC material for interesting angles/extra analysis; linking to BBC/non-BBC pieces; galleries), break this down into 'digestible' snappy entries and publish. Usually, we have at least two people running a live page: one would be writing/modifying entries fed by different outlets (writer); the other would be checking them and publishing (sub). But sometimes you run LP on your own.
>
> (Interviewee 102, BBC 'Core' writer)

The live page would work in conjunction with the UGC hub, particularly over video. The video would usually be verified by the hub, relevant BBC language services and the BBC's monitoring department at Caversham, before the live page would use it. The policy at the BBC was to 'refer up' any sensitive material, which may help to explain why so little video was used. The idea, however, was to create an interesting format for BBC viewers to look at.

The BBC live page was dominated by authority figures (72 references), with frequent use of quotes from the World Nuclear Association, President Obama, Hillary Clinton and senior British politicians such as David Cameron and William Hague. It also relied heavily on quoting its own journalists—around 45 of the sources could be described as BBC correspondents. Most of them were on assignment, although there were a couple who were described as eyewitnesses because they were on holiday there at the time. The BBC took many of its updates from the wires—63 references in total—but this is probably because of its one-source rule as, the core writer explained, "any one-source development is attributed to news wires (Reuters, AP, AFP etc.)".

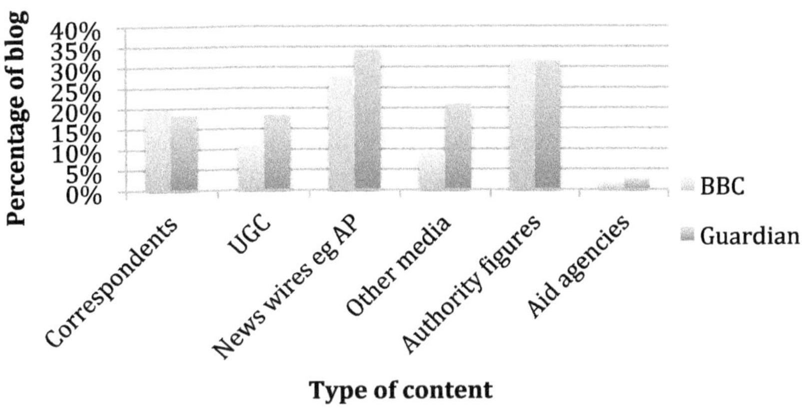

Figure 7.3 Types of Content on BBC and Guardian Liveblogs 11 March 2011

The BBC also credited other media—mainly the Japanese TV station NHK, but also outlets such as the *New York Times* and the *LA Times*—21 times.

As with the *Guardian*, there were few NGOs quoted (only four sources in total) but there was an effort to include UGC—which constituted around a tenth of the blog. This mainly came down to embedding tweets.

It is widely believed that the creators of user-generated content tend not to represent the public at large. For example, a typical UGC contributor to the BBC was a 45- to 54-year-old male, employed full-time as a middle manager or professional (Wardle and Williams, 2008). This view is challenged, however, by Bergstrom (2008), who found while younger, well-educated internet users were more likely to create content, those of lower or middle-level education were most likely to think it was important to comment on news websites.

It is difficult to draw conclusions from such a small number of comments on the liveblogs, but the *Guardian's* blog largely seems to back up Bergstrom's argument, while the BBC tends to vindicate Wardle and Williams.

The *Guardian*'s UCG creators I managed to contact were aged between mid-twenties and late-thirties. Unsurprisingly, most were expats—largely from England, although there was a Scot and a Chilean. Two of the content creators were English teachers, while a couple were IT specialists. There was also an engineer, a music journalist and a communications specialist. They were all based in Tokyo.

The BBC's UGC creators were more diverse, which probably reflects the fact that there were more contributors, as well as the BBC's greater global reach. They ranged in age from 20 to 50 and included a professor, two teachers, a trader, a diplomat, an actor, a student, a small business owner, a tourist guide, an IT specialist, a journalist and a philosopher. Most of them came from Tokyo, but because the BBC's blog ran longer than the *Guardian*'s, it also used three contributors from Hawaii, which was put on alert after the tsunami. As with the *Guardian*, the majority were white, although some declined to give their ethnic background.

There was a marked preponderance of male creators of UGC for both blogs, although it was impossible to tell the gender of the small number who did not respond to requests for interview from only their usernames or Twitter handles. Those who responded to requests tended to be male as well; all seven *Guardian* interviewees were male. The numbers, however, are too small to make firm judgements about the reasons for this.

Prior Warning of Content Being Used by Mainstream Media

There was a sharp but understandable divide in the reaction to hearing their contributions had been used in the mainstream media between those who had posted information up on social networking sites such as Twitter, and those who had gone on mainstream media sites to post messages themselves. Most of those I contacted had no idea that their information had been distributed to a wider audience in this way. Of those who answered this question, from the *Guardian* five had not been contacted while two had, and of the BBC, two had been contacted, and 11 had not.

> I didn't know that my content had been used in the BBC's liveblog. Can you send me what was used and/or how? (Obviously, this question answers some of your questions.)
>
> (Interviewee 63, journalist)

> I just noticed now that my tweet was used by *Guardian*. My tweets from those days were used by so many newspapers and TV in Spain that I lost track.
>
> (Interviewee 50, engineer)

Some became aware because the number of people following them on Twitter suddenly began to jump (in particular Interviewee 55, a small business owner, whose tweets were used on the BBC blog; Interviewee 54, a student, who saw his followers increase by 50 per hour; and Interviewee 48, the music writer on the *Guardian* blog who saw his rise by 100 an hour).

There was a fairly fatalistic attitude to the use of such content. While most did not expect it to be picked up by the mainstream media, those who had tweets used did not blame the media for using something that they themselves had posted in the public domain. For those who had directly commented on the *Guardian* webpages or taken the time to fill out a form on the BBC site there was, again, a lack of prior warning that their content was about to go up on the site. But they realised that by interacting with a media organisation in this way, their content was very likely to be used.

Interviewee 56, who was working for a technology company in Hawaii, was contacted the BBC, but his own background as a journalism graduate may have made him more open to journalists' requests:

> I was happy to contribute and, as an 'open news' kind of guy with a Creative Commons license on most of my media, I would've been fine with use with credit without prior approval. I was actually impressed anyone bothered to ask!

Only one creator—Interviewee 54 whose tweet was used on the BBC blog—raised questions about the fact that no one appeared to have tried to verify his words:

> I was a little bit surprised that the BBC would use a tweet of mine, because they didn't know me, and I was a little bit surprised because I asked myself how would they know I'm really there? Everyone can put in a location and say something but how did they check about the validity or the legitimacy of the tweet—that's what I didn't understand.

Ironically, there was most dissatisfaction when a journalist had reverted to more traditional methods to cover a story.

Interviewee 46, a digital media producer, had started tweeting immediately after the quake struck and his tweets were picked up by the *Guardian*. He was then contacted by a *Guardian* journalist, who interviewed him as he walked home on his iPhone via Skype (phone lines were down but data networks were still working) about his experience during the quake.

This, however, was written up as a first-person 'ghosted' piece—a traditional journalistic device where the journalist interviews a celebrity or an eyewitness to a particular event, then writes the story in the first person. Usually it is good practice to either allow the subject to approve the piece or put a disclaimer at the end of the article, such as 'As told to [name of journalist]'. Interviewee 46, however, claims that, while he was fully aware that he was being interviewed by a journalist, he was not told this would appear as a first-person piece.

> [The *Guardian* reported it] fairly accurately, but to be honest I didn't feel good about it. I felt that I was misled in a way, because I was. I'm not sure I was given the [right] impression or whatever, I just had assumed that it would be. . . . It's not too bad, it's not too embarrassing, but sometimes people, when they're talking, have a tendency to exaggerate, and especially there was so many appalling articles by foreigners saying, oh, we've got to escape, blah blah blah, just all these shock stories, and thankfully, although I was in shock and it was all happening just then it seems that it wasn't that inaccurate actually. It wasn't that they misreported, it's just that I felt they should have been upfront in how it was going to be presented.

The Virtual Doorstep

How journalists should approach those caught up in tragic events or disasters has been the subject of much debate (McLellan, 1999; Ewart, 2002; Townend, 2012; Newton and Duncan, 2012). The 'doorstep' and 'death knock' have frequently been jobs given to junior reporters because of journalists' dislike of such tasks (Cooper, 2012)—with the result that inexperienced young journalists are all too frequently sent to deal with bereaved and traumatised people.

Still, there have been considerable efforts to educate journalists in how to approach survivors and families. In the UK, Section 5 of the Independent Press Standards Organisation's Code[1] specifically deals with intrusion into grief and shock, as does Section 8.16 of Ofcom's Broadcasting Code Guidance.[2] Institutions like the Dart Center for Journalism and Trauma have produced their own suggestions.[3]

In the main, these forms of guidance have concentrated on journalists who meet their subjects face to face or deal with them over the phone. Increasingly common, however, is the problem of the 'virtual doorstep', where those who have been involved in traumatic events are pursued in cyberspace. The lack of consent in the unauthorised use of social media material is sometimes painful for people, because they see it as 'stealing their identity' (Newton and Duncan, 2012). Approaching people in cyberspace is often brief, abrupt and lacking the social niceties of checking how survivors or families feel in the aftermath of a shocking event. This has led to some journalists to call for more thought and guidelines as to how they should approach survivors of traumatic events via social media (Frankel, 2015; Kennedy, 2015).

While creators like Interviewee 56—who had some experience of journalism and saw himself as a champion of open media—did not mind the media attention, others whose tweets or content were used by the *Guardian* or the BBC often found themselves pursued by other media outlets as a result, and in many cases the 'virtual doorstep' proved stressful. They found the volume of interest and the personal interest in themselves particularly overwhelming; they often did not realise that the media was interested in who they were, as well as the material they had created.

Interviewee 51, a teacher, ended up deleting his Facebook account, while Interviewee 58 who worked in the tourist industry was contacted by between 15 and 20 journalists. He said they were polite and kind in the main, but when harassed by a BBC journalist who disputed his account that all was calm in his neighbourhood (see full story below) he had found it very stressful. So when another BBC reporter rang him the day after he refused to speak to them, saying: "What is the point of me answering to you if you don't believe me anyway?"

Interviewee 43, an IT manager, had filled in a *Have Your Say* form about his experience and was called by a BBC journalist. His recollection was that he was told he was about to be put on air "within a minute or two" without time to consider. This had adverse consequences for him.

> My company at the time are [sic] very media sensitive, so after the 'spot' interview, not only did I email our Corporate Communication team about having been interviewed on the air (albeit in my personal capacity, with neither my work or company being mentioned). Nonetheless, I was verbally told ('reprimanded'?) (sic) not to do any further discussion/interviews.

Interviewee 54 (the student) also had an unhappy experience. He was approached by the Munich-based newspaper *Suddeutsche Zeitung* and a German/Swiss television station as a result of his BBC tweet. These approaches were by phone and face to face but he found the way that his details were shared online difficult. He was about to go back home to Germany to try to get a job, so he did not want his full name used in the newspaper article or for the interview to be shown in Germany itself. In the event, while the newspaper journalist only used an abbreviation of his last name, the student could still find it by Googling himself. In addition, the Swiss journalists sold the interview to German channels and his family saw him.

> They didn't really value what I explicitly asked them—not give away some information or not broadcast in some region, but they did anyway. So I kind of have the feeling I cannot trust journalists anymore, because you never know what they are going to do with your stuff.

Another interviewee whose content was used was from Wales but had tweeted a general message of support: "Thoughts are with the people of Japan and all the neighbouring countries that will be affected by the tsunami". When it was used on the BBC blog, she found herself under attack. A blog reader complained about the use of tweets in news stories to the BBC. This complaint was then mentioned on the BBC News Channel's feedback programme, *Newswatch*, which used her tweet as an example and clearly displayed her username. As a result, having been unaware the BBC had even used her tweet, she started to get critical tweets from strangers.

> I didn't mind it being used although didn't really understand why. The tweet was very generic. When I got mentioned in the complaint about social media in the news the 'trolling' was a bit of a nuisance.
>
> (Interviewee 53, radio station worker)

There was an understandable reluctance on the part of the UGC creators I interviewed to acknowledge that money would ever play a part in their decision to share content; and some people found themselves attacked online because they were assumed to be making a profit out of it. For example,

Interviewee 65—the security guard who captured the 2013 Oklahoma hurricane on a Vine video—eventually took all his videos down after being criticised by other social media users for supposedly profiting from his content.

Going Viral: The Consequences

While those whose content was appropriated by the *Guardian* and BBC Great East Japan liveblogs found it annoying (or worse in the cases of Interviewees 43 and 53) there were more severe problems if an eyewitness took a potentially crucial piece of footage while the disaster was still unfolding. This was the case with Interviewee 65, whose Vine videos were seen as the key images of the 2013 tornado in Oklahoma, and Interviewee 66, whose picture of the Vauxhall helicopter crash of the same year made the front pages of the newspapers.

> For about a day and a half I was hounded by media attention. It was all pretty overwhelming to be honest. . . . For many days after, I was pretty stressed about what I had seen and felt guilty about my part in taking videos of it. I did not profit in any way from this and I have found it awkward in wanting to go back to using social media as I had before this event occurred. That's where my head has been lately, trying to get back to normal, but pretty mentally discouraged.
>
> (Interviewee 65)

Interviewee 65 was a security guard who lived about two miles from the tornado's epicentre. He posted a series of Vine videos (six-second videos which can be shared via Twitter) showing the devastation it had caused. In one, the tornado swept across the area in front of the camera; in another, men were seen searching for a little lost boy called Tommy. He uploaded the videos and started to drive home.

> After I uploaded the videos, it took maybe 20–30 minutes for the attention to become widespread. I became aware of a steady stream of RT's [retweets] during my ride back home. By the time I returned home from uploading them on the scene, it had been about 40 minutes. By the time I returned home, I was bombarded by Twitter mentions and was on the phone talking with CNN.

His videos were picked up widely by media outlets across the world, from Australia to the UK to the US (Telegraph, 2013; CNN, 2013;ABC News, 2013) and he found himself under considerable pressure online to respond to journalists' requests. Interviewee 65 found it "overwhelming" and "complicated", particularly when he was bombarded with requests from different journalists at the same network. He agreed to allow all journalists who asked permission to use his video, but one of his followers had reprimanded

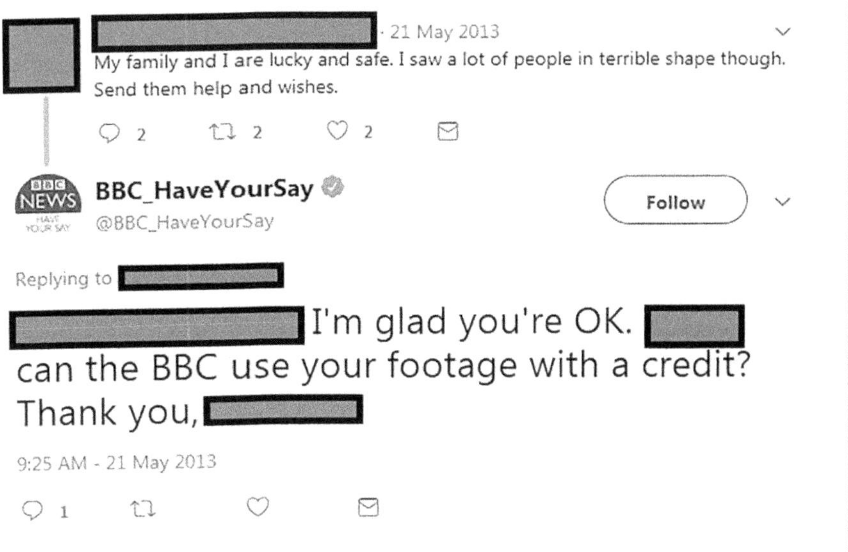

Figure 7.4 Interviewee 65 Conversation with BBC UGC Hub

Figure 7.5 Interviewee 65 Conversation with Swedish Newspaper *Dagens Nyheter*

Figure 7.6 Interviewee 65 Conversation with Local Stations News 10 (Sacramento, California) and 12 News (Arizona). A Political Activist Steps in to Remind Him He Can Make His Account Private

AP for not crediting him. He said he still found it "too painful" to look back over his Twitter feed, but was willing to share messages, of which these are a small selection.

His feed (provided by him) does show many of the journalists make brief reference to his own health and safety before going on to request permission

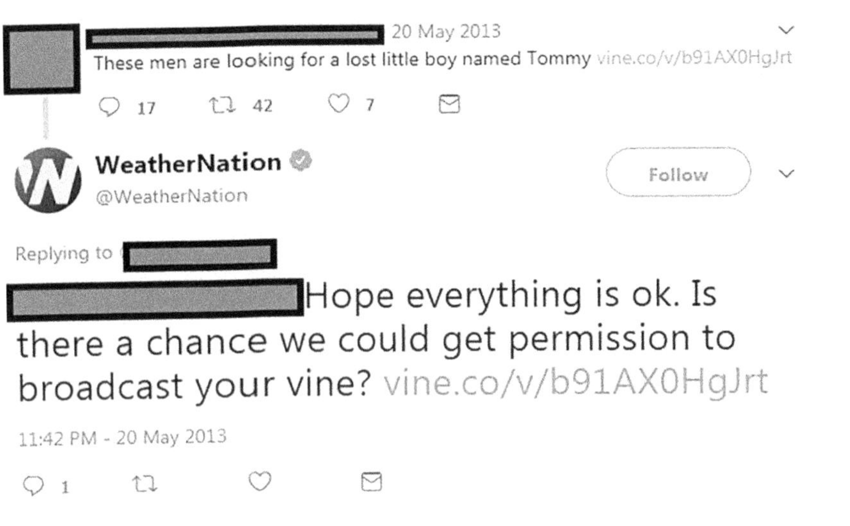

Figure 7.7 Interviewee 65's Conversation with Weather Nation

Figure 7.8 Interviewee 65's Conversation with the Huffington Post

to use the Vines. Others, however, are caught up in the 'story' and make no reference to what he may be feeling; they frame it as what Chouliaraki terms 'ecstatic' news—news where there is a break with the news conventions and a "move . . . to an uninterrupted flow of images and stories with

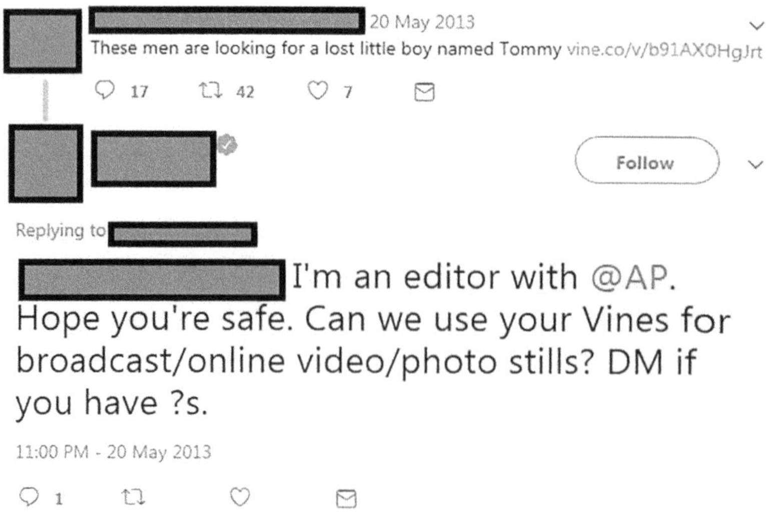

Figure 7.9 Interviewee 65's First Conversation with AP

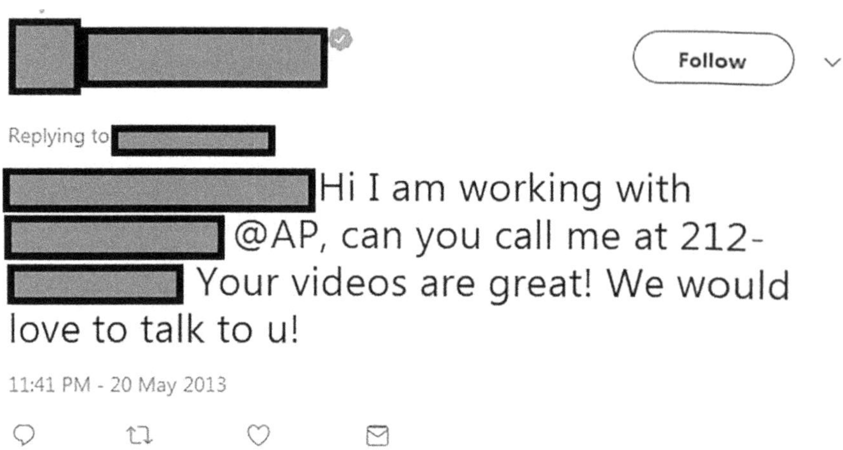

Figure 7.10 Interviewee 65's Second Conversation with AP

various degrees of emotional power" (Chouliaraki, 2008). Some journalists use adjectives such as "terrific" and "great". Interviewee 65 said that at no point did any journalist discuss payment or copyright with him, even though AP and Storyful would have been distributing his content to their subscribing clients.[4] Like many of those who contributed to the Great East Japan quake liveblog however, he felt incredibly uncomfortable about money being discussed.

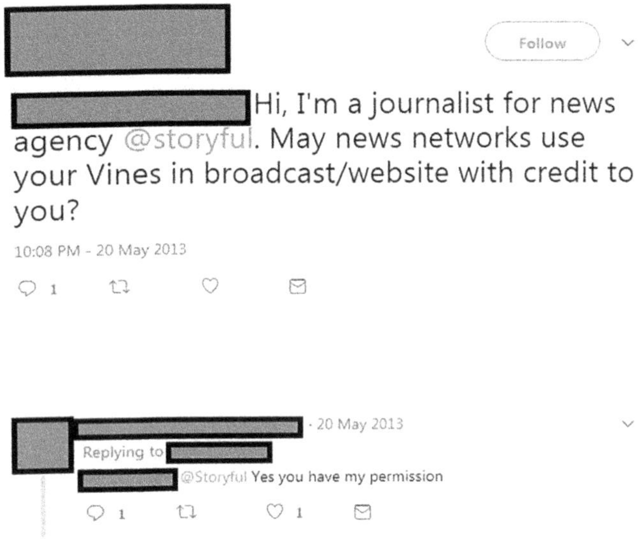

Figure 7.11 Interviewee 65's Conversation with Storyful

> Various friends, family, and acquaintances discussed with me in the days after that money could have been an option when giving permission for video use. As someone who has never been in this kind of scenario before, I was unaware that that was even an option. One of the hardest things for me mentally is getting over that I could have done more to help the people I had seen in that situation. I don't feel any sort of joy or want for attention in the benefit for myself because of what I had seen happen to those people. I couldn't live with myself to profit from others suffering. I was glad to share my videos and money was never an option discussed with anyone in any form.
>
> (Interview 65)

As with Interviewee 53, the virtual doorstep took two forms. While being pursued by journalists on Twitter caused Interviewee 65 stress and panic, his very visible engagement laid him open to criticism by internet trolls. He received many abusive messages from members of the public, mainly accusing him of wanting glory and failing to help the parents searching for a child.[5] In the end he deleted all his Vines and temporarily left Twitter because of the criticism he received for taking videos rather than helping with the relief effort. It was something that he felt unable to cope with and used words like "depressed" to describe himself.

> I was unequipped to know how to deal with such attention. Being famous is something we all may dream about, but when nationwide

> attention came to me, especially due to a natural disaster and not of my own accord, I wasn't happy about it. I would tell others to just follow their gut in how they should handle any situation. There is a weird push and pull when it comes to such attention. For me, I didn't want so much attention or even try to get some sort of personal gain after witnessing so many people in a bad situation. I did a few interviews and then stopped interacting because I needed to take time for myself. It didn't take them long to move on.

On 16 January 2013, Interviewee 66, a financial controller, had been about to leave for work in Stratford, east London, when a helicopter crashed in Vauxhall near where he lived; Interviewee 67, a video games designer, was still in bed and woken up by the sound of the helicopter and the subsequent crash. Both immediately rushed out to see what had happened and were some of the first eyewitnesses. Interviewee 66 went on to take some photos on his phone, while Interviewee 67 took video and some inadvertent stills.

At the suggestion of construction workers outside his home, Interviewee 66 tweeted his picture—it was the first tweet he had ever sent. Inexperienced in social media, he thought no more about it as he was in back-to-back meetings that day. He did not look at Twitter again until later that afternoon, when friends alerted him to the interest his picture was generating.

> I went to work pretty much oblivious to everything that was happening on Twitter until about 3 or 4 o'clock when one of my friends finally got hold of me. I had about three or four or five hundred emails mainly from journalists, saying, "can we use the photo?" "can you get hold of me?" And some other friends who had seen it on the news, but I had no idea. In one of the messages was one of my friends saying, "oh your picture's on the front page of the Evening Standard", and that was the first I'd heard about it.

Unlike with Interviewee 65, whose name was generally used in conjunction with his Vines, Interviewee 66 discovered that as well as the *Standard*, his picture had been used by Swedish, Danish and New York newspapers, without any credit or recognition that it was his work. When used by BBC TV, they attached his Twitter handle. He later became the subject of a *Guardian* article investigating the way that mainstream media were failing to respect copyright or labelling when using UGC (O'Carroll, 2013). He was less concerned about copyright or money; he felt that he had put it up on the net and it was there for people to use. Like Interviewee 65 he felt extremely awkward at the thought of 'profiting' from someone else's disaster.

Interviewee 67, the video games designer, meanwhile, had been outside and even approached a burning car with another passer-by to check there was no one left inside; he retreated when paramedics and police arrived and it was clear there was nothing he could do to help. An early Tweeter (on

Figure 7.12 Interviewee 67's narrative (reads chronologically from bottom to top)

Figure 7.13 Interviewee 67's Tweet Being Shown on Sky News

the site from March 2007 and among its first million users), he instinctively tweeted stills and a narrative of what he saw, as well as posting the video on YouTube. He had not even finished uploading the video when he saw his tweet being used on Sky News.

His tweets were used by very many media outlets—ranging from heavyweights such as the BBC, AFP, AP and Reuters to more obscure ones such as Belarussian television and Scandinavian outlets. He asked for credit and said in the main he got it—there were a few improper uses but not many.

> I was fine with it. The context was that I'd just had a major disaster happen on my doorstep. I'd been removed from my home and was handling media queries via an iPhone in a cafe. The thing I was concerned about at this time was getting the news out there. It was about timeliness and helping news media define the incident by using first-hand material. What I *wasn't* considering was being paid for it.

Perhaps because his content included video as well as stills, Interviewee 67 said that he was asked by people after the event how much money he had made from it. He said that no journalist had ever offered to pay for his content; the requests had only ever been for permission to use it.

> I never considered money at the time, it was probably an hour or two after I'd been dealing with the incoming queries that other participants on Twitter started to get fruity about it. One Twitter user told me to take it down—he represented a broker and would monetise it on my behalf and give me 50%. I could also have turned on monetisation on the YouTube video. It's worth bearing in mind that no matter how the media organisations were wrapping this video, the unedited source was available on YouTube. 4.5 million people watched it but there was no way I could stomach monetising a video in which, I figured, a number of people had died. It just seemed like an awful thing to turn into a profit opportunity.

He did feel, however, that—while it was unlikely that eyewitnesses would want to monetise their content—journalists should be more open about how they deal with eyewitnesses, perhaps offering compensation afterwards or providing a pro-forma that would explain that they were rescinding their rights.

Where both did feel strongly was about the way journalists had approached them in cyberspace. Interviewee 66 estimated that by mid-afternoon he had had hundreds of messages via Twitter and email, which he found overwhelming. His main concern was that it was not only his 'virtual doorstep' that was being over-run by journalists.

> I'm quite aware that everyone can see what you post on Twitter, not that I tweet, but if I did. I think it's different for Facebook, because

> Facebook is for friends and there should be some sort of privacy on that. The only issue that there has been on the privacy front is that there were a number of people who tried to contact me through there [Facebook and Twitter] and that it was pretty relentless. I don't really mind saying it was pretty stressful because I had this guy from the BBC who was harassing my followers on Twitter to try to get hold of me and he didn't message or email me but went for them. There were a few work people, like our Commercial Director, who got contacted. Not that it's the worst thing in the world, but it was like "I don't really like this feeling".

He also felt strongly that the tone used by the journalists on email and Twitter was often rude and abrupt, dispensing with social niceties. While Interviewee 65, the security guard, felt that most journalists had at least attempted "mandatory pleasantries", for Interviewee 66 the reporters' approaches had been aggressive, coupled with the sheer scale of interest. Interviewee 67 talked about "emotionless" contact on Twitter and email, and the "persistence" of TV crews approaching him via cyberspace. He declined all TV interviews because he felt shaky, although he did two print interviews with the *Telegraph* and the *Standard*, which he described as almost therapeutic.

While both Interviewees 66 and 67 say they would share content again if they found themselves in the same situation, they felt strongly that journalists' conduct needed to change.

> I think if I *had* been in shock, or lost someone, or my home had been damaged etc., these messages would have a very different impact on me.
>
> (Interviewee 67, video games designer)

> I know their [journalists'] job is really time pressured but for people who aren't used to dealing with that it can be pretty daunting. . . . If you are tweeting someone it's easy to forget there's a real person, and that came across. Some of them were really rude '[I] need to speak to you immediately'—it was very single sentence emails, and a lot of them.
>
> (Interviewee 66, financial controller)

"I Had to Take a Serious Responsibility": Eyewitnesses or Journalists?

> As I said to my staff, officemates, friends and anybody who cared to ask at the time, "Facebook saves the world". Since cellular communications were effectively cut off (majority of bandwidth in use by government/emergency services), only IP-based communications (i.e. the 'internet') was available.
>
> (Interviewee 44, IT manager)

Were these tweeters citizen journalists? None of them identified themselves as such; they saw themselves as witnesses who were caught up in a disaster.

But what was revealed was that those whose work was used by the mainstream media (once they were aware of it) started to change their behaviour as a result. Some felt they had to start thinking much more about the tweets they were publishing, out of a sense that they now had a public role, rather than simply making private comments.

> I felt it was wise to contribute at the time as people wanted to know what was happening. It's embarrassing looking back on it but the air raid sirens and tsunami warnings were blaring for 6–12 hours where I lived and my area was also due a massive earthquake.
>
> (Interviewee 47, English teacher)

For some there was an empathetic and altruistic role to be played. Interviewee 61, who worked at an international school, said he had felt "grateful" he had been able to reassure families abroad that their loved ones were safe. Interviewee 59, a professor at Hiroshima International University, described his tweets as a "mission" and "self-relief"; he felt that this was practical help he could offer.

One of the most interesting tweeters was Interviewee 64, a Japanese trader who had been tweeting in both Japanese and English. She had been stuck in her office when the quake hit. Originally she had started tweeting as a way of communicating with other people caught up in the earthquake—so she and her friends could find out where the damaged areas were. Tweets were also a way that people who were left isolated, often on apartment rooftops, could be traced and rescued by friends, government agencies or the Japan self-defence force (the military). "[That] information moved people to feel they would like to be a 'HELP' for any reason". She was unusual in that she also tweeted in English, primarily because she had friends from abroad as well as in Japan trying to find out what was going on.

> I thought my tweet information might help someone because many Japanese people won't tweet in English but I could. . . . As you can guess, suddenly my tweets were retweeted by many people who were not my followers . . . I got many foreign followers in a few days that made me feel that I should tweet about the earthquake in English for those new people. Through this experience I learned I had to take a serious responsibility about all my words in Twitter even [though] I was just [a] common citizen in Japan.

These tweeters did not identify themselves as citizen journalists, but they saw themselves as fulfilling some kind of public service. They were interacting with their followers not just to pass on emergency information to those caught up in the quake, but also to address a wider audience that had been led to them by the mainstream media.

But does this take them beyond an eyewitness role? In analysing Sohaib Athar's (@ReallyVirtual) tweets of the raid which killed Osama bin Laden, Myers (2011) makes a persuasive argument that those who do not identify as citizen or accidental journalists may still be participating in "acts of journalism" (Stearns, 2013). Rather than simply telling friends what has happened, they find themselves caught up in a newsworthy event and start to act in a journalistic manner. Crucially this is partly because eyewitnesses are now publishers too—although where they think they are publishing and the act of 're-publishing' by mainstream media is frequently where conflict arises.

In Athar's case, he observed something unusual and shared it, answered questions, tried to act as a conduit for information and sought corroboration. This does not make him a professional journalist, but Myers argues it does mean that he is acting journalistically. As he puts it: "An amateur slugger probably won't hit a home run off a major league pitcher, but they're both playing baseball" (Myers, 2011).

Many of those who were caught up in the quake did nothing more than bear witness—writing a tweet, sharing a picture. But several did start to go beyond that and not only share information, but also correct information that was out there.

Interviewee 48 was particularly concerned by what he saw as tweeters tweeting what journalists expected to hear—and sensationalising what was going on (this was in the aftermath of the quake, but before the impact of the damage to the nuclear facility at Fukushima became clear).

> Suddenly prolific tweeters became 'eyes on the ground' and they tended to 'act' the way they suspected a journalist should act. Another guy ended up on [the] BBC soon after the quake talking about his experience—but he was in Nagoya, which couldn't have possibly felt the quake that strongly as it's very far from Tohoku. I thought those citizen journalists were telling the newsreaders what they suspected the public wanted to hear, and thus you got a lot of incorrect information overseas.

To counter this and put the relative strength of the quake in different parts of Japan into perspective, he posted a picture of a milk carton that was on the edge of his sink and did not fall off during the earthquake. The picture later became the subject of a Pecha Kucha Night in Berlin.[6]

Interviewee 58, a Russian working in tourism industry, had similar difficulties. He had started to tweet mainly to reassure his parents in Russia, who were becoming increasingly worried about scare stories in the Russian media about what was happening in Japan. As a result he estimated that around 15–20 media outlets contacted him. He gave them information and personal anecdotes to illustrate what he termed real—not 'made up'—problems. He said that he had found the vast majority of journalists respectful and kind, but he had had difficulty persuading one BBC journalist that the situation on the ground was not as bad as she wanted it to be.

> I lived in Niigata, a prefecture next to Fukushima prefecture. A regular life was continuing in Niigata—people didn't stop going to their job, children were going to school as usual, women were taking their small children to kindergartens. People were collecting donations at the city stadium (and there were really many people bringing needed things and really many volunteers were helping), but no signs of panic. I was really impressed by the calmness of the Japanese people and their cooperation without any complaints about the situation and I talked to that BBC journalist about that. But she answered, "Sorry, I cannot believe you. We have information that there is panic, people have lack of food and electricity. . . " I told her that I am talking not about Fukushima or Miyagi (in those areas it was a really bad situation), I am talking about Niigata where I live, it is 200km away from Fukushima. . . . But she couldn't believe me: "The situation is awful around all Japan. . . . Sorry, but I cannot believe you. Could you give me a contact information of some other people living in Japan?"

So when the BBC called him the next day, he initially refused to talk to them because of the journalist's refusal to believe him. It was only after the second journalist promised to look into the matter and talk to the first journalist that he agreed to another interview, because he wanted to ensure that the record was set straight. Interviewee 59, a professor also refused interviews from US radio stations because he felt Hiroshima was too far from the epicentre and thus not part of the situation journalists wanted to report.

Taking Back Control

That is not to say there was no attempt to take back control. A German student studying for an MA in Tokyo, whose tweet was used by the BBC, said that he had sought information from other sources (Interviewee 54). These included Michael Gakuran's blog (gakuranman.com)[7] and a blogger who used the pseudonym Our Man in Abiko, who collected experiences of people involved in the quake via the Twitter hashtag #quakebook.

> [Michael Gakuran] covered this whole incident not from a journalist point of view but as a journalist wannabe. I don't know him personally but many people looked at his website during these days to gather information. . . . He collected everything, and made a better timeline and a more informative one than in the mainstream media.
>
> (Interviewee 54)

Gakuran, a British web designer and marketer, had previously used his blog to document his experiences of living in Japan and to showcase his photography. He lived in Nagoya, some way from the epicentre of the quake, but saw friends from Tokyo and Kyoto on Twitter reacting to the disaster at the same time. His memory of 11 March was of "Twitter going crazy with people tweeting bits and pieces of information, the chill of watching the tsunami destroy villages and kill people, live online".

Concerned also about what he saw as the 'apathy' of his co-workers towards those who were suffering, Gakuran started to curate information sent to him, or that he found in his Twitter stream (he had followed a large number of foreign nationals living in Japan for some time). Crucially, much like other UGC creators he started to try to clarify information and distribute it during the month after the quake. Like Interviewee 64, the Japanese trader, he also mentioned (unprompted) a sense of "responsibility" to his audience.

> I wanted to make sense of everything that was happening and also document this historic event. . . . Increasingly, though, as more and more people followed my Twitter account for information, I felt a sense of responsibility to provide a cohesive, unbiased account of everything and help people who needed information, so the posts took the form of livestream news updates (now quite commonplace for big media outlets when breaking news occurs) as well as a hub of links to sites which offered help to those in need.

Gakuran saw his main audience as English speakers living in Japan, so one of the aims of the blog was to translate Japanese information. Another overt intention was to counter the nature of foreign press reporting, although it is interesting that even then he does not define what he is doing as journalism.

> The efforts of myself and many, many other writers online stood in stark contrast to the sensational reports from most of the foreign press, delivered a more balanced picture of the actual situation in Japan and, I hope, enabled people to make more informed decisions for themselves. Done differently, perhaps with more time I would have gone out to do *some real journalism* [my italics] on the streets to get a better picture of how the Japanese public felt and also put more effort into organising and structuring the posts. I would also have invested in a more robust platform as my site went down a number of times due to the surge in traffic.

Gakuran also said that his blog contributed to Quakebook, set up by a blogger known only as Our Man in Abiko (OMIA). OMIA declined my invitation to be interviewed, but sent a number of previous interviews he had done to explain the thinking behind Quakebook.

OMIA was working as an English teacher[8] and, like Gakuran, felt the need to do something to help. His idea, however, was to use Twitter to find personal accounts as a basis of explaining the experience of the quake, and give a voice to different people. Within a week he had compiled a book—Quakebook—consisting of 89 contributors, crowdsourced via Twitter:

> It started out as just a crazy idea. I was having a shower one week after the earthquake, and I thought. . . "What can I do to help? I'm just an English teacher, uh, in Abiko, uh, on the outskirts of Tokyo. Um, so what can I do?" . . . I thought I would collect people's experiences of the earthquake, put it together, sell the book, and all the profits will go to the Red Cross. That's the basic idea, and it's just grown from there.
>
> (Our Man in Abiko, 2011a)[9]

What was particularly interesting about OMIA's approach was that he swiftly realised that his contributors tended to be foreign nationals because he himself wrote in English:

> So my wife got in touch with people from some of the really hard hit areas, um, including Sendai, um, where there was a, a, a family who, uh, had to be evacuated, um, and they had a very, uh, harrowing story. Um, also, uh, near Fukushima. So, we had some people, not a lot, but some people who really, really did suffer a lot, and just gave their point of view, and, uh, the idea was to pull everything together, so no one person's experience is better or worse than another's. It just tells the story of what happened to this country, and the world, uh, on 3/11. That was the idea.
>
> (Our Man In Abiko, 2011a)

Despite being a former journalist, OMIA did not see his blog as journalism—he saw it as a historical record of the disaster and a way to fundraise. He also invested it with a moral purpose. In an article for the *Japan Echo*, he linked the decision to write Quakebook as a deliberate and normative alternative to leaving the country in the aftermath of the earthquake:

> More importantly to me, at least, I hope we have begun on our journey to pass the test that *Quakebook* contributor Tabe Yūshi believed the earthquake forces us to take: "Isn't this test for us to see whether or not we can start to become Good People?" . . . It's a test we must pass.
>
> (Our Man In Abiko, 2011b)

Conclusion

It has been more than a decade since members of the public first began providing defining images of major disasters, starting with the Indian Ocean tsunami (Cooper, 2007). Like other sources, the testimony of ordinary citizens has been an authenticating and emotive voice when used as part of the mainstream media narrative. It has provided information and images that journalists themselves could not have done (Glocer, 2006; Allan, 2009) and occasionally changed the way the stories were framed (Robinson, 2009).

In the fast-moving news cycle, the mainstream media's appropriation of content may happen before creators are even aware of it. This chapter investigated whose content is being used and what these creators felt about (sometimes unwitting) participation in the process of reporting humanitarian disasters.

The quantitative part of the chapter finds that UGC is now an established element of the mainstream media's coverage of humanitarian disasters. But the idea that more diverse voices are heard—or that technology has brought about a democratisation of humanitarian reporting, acting as a catalyst for a power shift—is overly optimistic when it comes to liveblogs. While both blogs did use a significant amount of UGC from ordinary people, it was far less than from authority figures and organisations. These people and organisations have learned to use social media effectively, in the process becoming what I call 'primary Twifiners'—adapting Hall's term to show how, even in a world where anyone can publish, journalists still favour authority sources.

The mainstream media tended to use content from generally well-educated expats in middle-class occupations. The sample was too small to make definitive conclusions but, like in other research (Wardle and Williams, 2008), was male-dominated. Media outlets rarely sought permission for using these content creators' text, but were more likely to contact them if they uploaded photographs and film.

Most of the 'accidental journalists' were happy, even flattered, that someone was interested in their content; there was a general acceptance that putting something on the internet meant that the mainstream media might well use it. There was a lack of knowledge about permissions and copyright and, because of the nature of the stories, there was also a squeamishness about seeing their content as something that could be monetised.

Many had no knowledge, until I contacted them, that their content—generally a short tweet—had been used by the mainstream media. So there were no adverse effects for them in terms of harassment or a loss of privacy. For those who were alerted by friends or family that they had been singled out, there was a change in attitude. What had started as often a shock response, a need to let family and friends know they were safe, or even a therapeutic act (Liu et al, 2009), became a chance to perform a type of public service. They did not see themselves as journalists or 'citizen journalists' but as going beyond witnessing to providing useful information to their followers. Some also wanted to act as a corrective, if necessary, to other media reports (in particular Interviewees 48 and 58). There was a strong sense, however, that others had been tempted to fall into line with the mainstream media's narrative. Conversely, when tweeters and bloggers tried to challenge this narrative, the mainstream media reacted with disbelief and dismissal.

Those whose content was sufficiently interesting to see them pursued by journalists were often surprised that these journalists were interested in finding out about them personally, and not just what they had witnessed. While they had been happy to share information via their own Twitter feeds or Facebook accounts, they found it unnerving to find themselves splashed across a newspaper or website (in particular Interviewees 46 and 54); a difference the journalists may not always have been able to see.

For those pursued in cyberspace, it was often a distressing and overwhelming experience. The compressed nature of a tweet (140 characters) or the speed with which a curt one-line email could be dispatched meant that those like Interviewees 65 and 66—who had particularly valuable images for the media—were often hounded, not just by many different media organisations, but by multiple reporters within the same organisations. Often this pursuit was handled with scant concern for how they might have been feeling in the aftermath of witnessing a dramatic event. It is easy for journalists to forget that they have become "silent watchers" (Smith cited in Cooper, 2012), seeing social networking sites as another valuable source and creators of content as another technological gadget, rather than people that—in a former journalistic age—they would have had to persuade face to face on a doorstep to hand over valuable images or text. As such the next chapter looks in more detail at the ethical issues that journalists (and aid agencies acting in the media arena) still have to resolve around this area.

Notes

1. See https://www.ipso.co.uk/editors-code-of-practice/. IPSO's predecessor, the Press Complaints Commission was in place at the time of most of the interviews for this book. Its code of practice can be found at www.pcc.org.uk/cop/practice.html
2. See http://stakeholders.ofcom.org.uk/binaries/broadcast/guidance/831193/section8.pdf
3. See http://dartcenter.org/topic/interviewing-victims
4. Storyful says on its web page that it does not 'sell' content, nor get a commission for content. "Our role in rights management is to surface newsworthy social content to news organizations, and inform our clients of clearance status of the content, and provide contact and credit information so that they can communicate directly with content owners if necessary to make payment." See 'Do you make money selling other people's content?' https://storyful.com/faq/
5. Typical messages from the trolls that Interviewee 65 received included abuse such as "Your vines did nothing you piece of shit. Next time tuck in your disaster boner, put down your iPhone, and help someone"; "Enjoy your five minutes of meaningless fame. Hope you're happy" and "Should you be tweeting and not helping people you ass-hole?"
6. Pecha Kucha Nights began in Japan in 2003. They are events at which young designers come to meet and share their work, and are characterised by swift presentations of around six minutes, with slides shown rapidly. They include creative work, research projects and other subjects.
7. The relevant entries can be found at http://gakuran.com/great-tohoku-earthquake-1/.

8. He had previously worked as a newspaper reporter and sub-editor; in his Amazon author page he refers to stints on the *Birmingham Post*, *Nottingham Evening Post* and *Derby Evening Telegraph*, before deciding to change career teaching English in Japan.
9. OMIA discusses on the blog his decision to keep in verbal tics as it is a literal transcript of an interview. I have also kept these in as it was a deliberate decision of his.

References

ABC News, 2013, 20 May-latest update *Aftermath of tornado in Moore* [Homepage of ABC News. Australia], [Online]. Available: www.abc.net.au/news/2013-05-21/aftermath-of-tornado-in-moore/4702198 [2018, 27 August].

Allan, S. 2013, *Citizen witnessing: Revisioning journalism in times of crisis*, Polity Press, Cambridge, UK.

Allan, S. & Thorsen, E. (eds). 2009, *Citizen journalism: Global perspectives*, Peter Lang, New York.

Bergström, A. 2008, "The reluctant audience: Online participation in the Swedish journalistic context", *Westminster Papers in Communication and Culture*, vol. 5, no. 2, pp. 60–80.

Borger, M., van Hoof, A. & Sanders, J. 2016, "Expecting reciprocity: Towards a model of the participants' perspective on participatory journalism", *New Media & Society*, vol 18, no 5, pp. 708–725.

Bruns, A. 2008, *3.1. The active audience: Transforming journalism from gatekeeping to gatewatching*. Available: http://snurb.info/files/The%20Active%20Audience.pdf [2018, 2 July].

Chouliaraki, L. 2008, 27 February, *Distant suffering in the media: Inaugural lecture at the LSE*, LSE, London. Available: http://richmedia.lse.ac.uk/publicLecturesAndEvents/20080227_1830_distantSufferingInTheMedia.mp3 [2018, 2 July].

CNN.Com. 2013, 20 May-latest update *Vine videos show Oklahoma tornado devastation* [Homepage of CNN.com], [Online]. Available: http://situationroom.blogs.cnn.com/2013/05/20/vine-videos-show-oklahoma-tornado-devastation/. [2018, 27 August]

Cooper, G. 2007, *Anyone here survived a wave, speak English and got a mobile? Aid agencies, the media and reporting disasters since the tsunami*, Nuffield College, Oxford.

Cooper, G. 2012, "Facing up to the ethical uses of Facebook", in *The phone hacking scandal: Journalism on trial*, eds. R. Keeble & J. Mair, Abramis, Bury St. Edmunds, pp. 353–230.

Ewart, J. 2002, *Prudence not prurience: A framework for journalists reporting disasters*, Proceedings of the 23rd Annual Conference of the Australia and New ZealandCommunication Association (ANZCA 2002). Available: www.researchgate.net/profile/Jacqui_Ewart/publication/228793152_Prudence_not_prurience_A_framework_for_journalists_reporting_disasters/links/02e7e528eb8acc8f70000000/Prudence-not-prurience-A-framework-for-journalists-reporting-disasters.pdf [2018, 2 July].

Frankel, M. 2015, 22 July-last update, *Journalists on Twitter: Stop shouting, start listening* [Homepage of BBC], [Online]. Available: www.bbc.co.uk/blogs/collegeofjournalism/entries/2c77f67e-b63f-4819-9a94-40c3bbff94f4 [2015, 12 August].

Gans, H.J. 2004, 1979, *Deciding what's news: A study of CBS evening news, NBC nightly news, Newsweek and Time*, 25th anniversary edn, Northwestern University Press, Evanston, IL.

Glocer, T. 2006, 11 October-last update, *We media speech* [Homepage of Tom Glocer], [Online]. Available: http://tomglocer.com/blogs/sample_weblog/archive/2006/10/11/98.aspx [2012, 20 May].

Harrison, J. 2010, "User-generated content and gatekeeping at the BBC hub", *Journalism Studies*, vol. 11, no. 2, pp. 243–256.

Hermida, A. & Thurman, N. 2008, "A clash of cultures", *Journalism Practice*, vol. 2, no. 3, pp. 343–356.

Hujanen, J. & Pietikäinen, S. 2004, "Interactive uses of journalism: Crossing between technological potential and young people's news-using practices", *New Media & Society*, vol. 6, no. 3, pp. 383–401.

Kennedy, R. 2015, 27 July-last update, *Consider the source: The importance of the eyewitness in real time news coverage* [Homepage of First Draft], [Online]. Available: https://medium.com/1st-draft/consider-the-source-106253680a59 [2015, 12 August].

Larsson, A.O. 2011, "Interactive to me—interactive to you? A study of use and appreciation of interactivity on Swedish newspaper websites", *New Media & Society*, vol. 13, no. 7, pp. 1180–1197.

Liu, S.B., Palen, L., Sutton, J., Hughes, A.L. & Vieweg, S. 2009, "Citizen photojournalism during crisis events", in *Citizen journalism: Global perspectives*, eds. S. Allan & E. Thorsen, Peter Lang, New York, pp. 43–63.

McLellan, T. 1999, "Fair game or fair go? Impact of news reporting on victims and survivors of traumatic events", *Asia Pacific Media Educator*, vol. 1, no. 7, pp. 53–73.

Myers, S. 2011, 24 November-last update, *Why the man who tweeted the Osama bin Laden raid is a citizen journalist* [Homepage of Poynter], [Online]. Available: www.poynter.org/news/mediawire/131135/why-the-man-who-tweeted-bin-laden-raid-is-a-citizen-journalist/ [2015, 12 August].

Newton, J. & Duncan, S. 2012, "Journalists and the bereaved: Constructing a positive approach to the teaching of death reporting", *Journalism Education*, vol. 1, no. 2, pp. 59–67.

O'Carroll, L., 2013, 16 January-last update *Twitter pictures put in spotlight following London helicopter crash* [Homepage of The Guardian] [Online]. Available: www.theguardian.com/technology/2013/jan/16/twitter-pictures-london-helicopter-crash-copyright [2018, 27 August].

O'Mahony, K. 2014, 25 January-last update, *As it happens: How live news blogs work and their future* [Homepage of Polis], [Online]. Available: www.lse.ac.uk/media@lse/Polis/documents/As-it-happens.pdf [2014, 1 July].

Our Man In Abiko, 2011a. '#Quakebook intercepted non-redacted cable.' *Our Man in Abiko*. Available at: http://www.ourmaninabiko.com/2011/09/quakebook-intercepted-non-redacted.html [Accessed: 27 March, 2016]. Sent via personal correspondence 22 January 2015.

Our Man In Abiko (2011b). 'Fight or Flight? The Tests We Face' Japan Echo. Available from: http://www.japanecho.net/society/0090/ [Accessed 27 March, 2016]. Sent via personal correspondence 15 August 2015.

Paulussen, S. & Ugille, P. 2008, "User-generated content in the newsroom: Professional and organisational constraints on participatory journalism", *Westminster Papers in Communications and Culture*, vol. 5, no. 2, p. 24.

Robinson, S. 2009, August, "'If you had been with us': Mainstream press and citizen journalists jockey for authority over the collective memory of Hurricane Katrina", *New Media and Society*, no. 11, pp. 795–814.

Robinson, S. 2010, "Traditionalists vs. convergers textual privilege, boundary work, and the journalist—audience relationship in the commenting policies of online news sites", *Convergence: The International Journal of Research into New Media Technologies*, vol. 16, no. 1, pp. 125–143.

Rosen, J. 2006, 27 June-last update, *The people formerly known as the audience* [Homepage of PressThink], [Online]. Available: http://archive.pressthink.org/2006/06/27/ppl_frmr.html [2015, 11 August].

Schlesinger, P. 1978, *Putting reality together*, Constable, London.

Singer, J.B. 2010, "Quality control: Perceived effects of user-generated content on newsroom norms, values and routines", *Journalism Practice*, vol. 4, no. 2, pp. 127–142.

Singer, J.B. & Ashman, I. 2009, "'Comment is free, but facts are sacred': User-generated content and ethical constructs at the Guardian", *Journal of Mass Media Ethics*, vol. 24, no. 1, pp. 3–21.

Stearns, J. 2013, October-last update, *Acts of journalism: Defining press freedom in the digital age* [Homepage of Ethical Journalism Network], [Online]. Available: http://ethicaljournalismnetwork.org/assets/docs/062/150/329303e-0d75796.pdf [2015, 12 August].

Telegraph, 2013, 21 May-last update *Oklahoma City Tornado: eyewitness films devastation of neighbourhood* [Homepage of Daily Telegraph], Online. Available: http://www.telegraph.co.uk/news/worldnews/northamerica/usa/10070355/Oklahoma-City-tornado-eyewitness-films-devastation-of-neighbourhood.html [2018, 27 August].

Thurman, N. & Walters, A. 2013, 12, "Live blogging—digital journalism's pivotal platform? A case study of the production, consumption and form of live blogs at Guardian.co.uk", *Digital Journalism*, vol. 1, no. 1, pp. 82–101.

Townend, J. 2012, 9 October-last update, *A journalist calls: Reporting death in a post-Leveson climate* [Homepage of Dart Center for Journalism and Trauma], [Online]. Available: http://dartcenter.org/content/journalist-calls-reporting-death-in-post-leveson-climate#.Vct86Uv_Qds [2015, 12 August].

Tran, M. 2014, 8 September-last update, *Royal baby announcement—as it happened* [Homepage of The Guardian], [Online]. Available: www.theguardian.com/uk-news/live/2014/sep/08/royal-baby-kate-and-william-are-expecting-their-second-child-live-updates [2016, 19 April].

Wahl-Jorgensen, K., Williams, A. & Wardle, C. 2010, "Audience views on user-generated content: Exploring the value of news from the bottom up", *Northern Lights: Film & Media Studies Yearbook*, vol. 8, no. 1, pp. 177–194.

Wardle, C., Dubberley, S. & Brown, P. 2014, *Amateur footage: A global study of user-generated content in TV and online news output*, Tow Center for Digital Journalism, New York.

Wardle, C. & Williams, A. 2008, "ugc@BBC:Understanding its impact on contributors, non-contributors and BBC news", *BBC*. Available: www.bbc.co.uk/blogs/knowledgeexchange/cardiffone.pdf. [2018, 27 August].

8 Ethical Questions Going Forward

Payment, Permission, Privacy

For those caught up in crisis events, there can be a real sense of satisfaction in being able to broadcast their thoughts or images to the world, a sense of connection in difficult circumstances, a validation of what they are going through. As we saw in the previous chapter, those using social media can often see it as a benefit to others and a means of helping in crisis situations. For journalists too, it may be the only means of getting footage or stills of an event as it happened—pictures or text that is "more newsworthy than professional content" (Hermida and Thurman, 2008:344). As Tom Glocer, the former head of Reuters pointed out, on 26 December 2004 none of Reuters' 2,300 journalists or 1,000 stringers were on the beaches the moment the wave struck. "For the first 24 hours", he said, "the best and the only photos and video came from tourists armed with telephones, digital cameras and camcorders. And if you didn't have those pictures, you weren't on the story" (Glocer, 2006).

Yet the mainstream media has often taken a smash-and-grab approach to such content—trawling Facebook, Twitter and YouTube and now Periscope and Snapchat for content, often without thought for the consequences. But this new approach to disaster reporting raises important issues around the rights and responsibilities for the people who create it and the media outlets that distribute more widely—often more widely than the creator ever thought possible. The difficult question of survivors understanding the nature of consent in the aftermath of disasters has been explored recently (Muller, 2013). But while most of this research relates to face-to-face interviews, I would argue it is equally applicable to online encounters as well.

Coverage of humanitarian disasters has always had complex ethical implications—Moeller's moral "bellwether" (1999) signalling normative responses on behalf of the audience. But in a current media environment which challenges our understanding of who can be a journalist, and how content can be easily shared and appropriated in ways far beyond what was originally intended, the ethics of reporting humanitarian disasters has to be addressed.

Ward (2005, 2008) argues that the globalisation of news media requires a rethinking of journalism ethics to a more 'humanistic' approach, a point taken up by Wahl-Jorgensen and Pantti (2013) who talk of the need of the journalist to be "actively involved in shaping and constructing the meaning of particular events in relation to understandings of the ethical responsibilities of the global community, nations and individual citizens/audience members" (2013:193).

So this chapter will look at payment (issues around copyright), permission (issues around consent) and privacy. It will also look at taste and decency. While some of these have their basis in legal protection, often (in areas such as privacy and consent) the legal, regulatory and ethical are closely bound up.

Different countries have different legal regimes but these are universal issues. This chapter will focus on UK and European examples, using as a basis the European Convention on Human Rights and Article 8 (which states that everyone has the right to respect for private and family life, home and correspondence), Article 1 Protocol 1 (which states that everyone has the right to peaceful enjoyment of possessions) and Article 10 (which provides the right to freedom of expression). But this approach is meant to facilitate debate rather than serve as a comprehensive legal summary.

Payment—UGC and Intellectual Property Rights

When a severe lake-effect storm hit Buffalo, New York in November 2014, Buffalo residents took to Facebook, Twitter, Instagram and YouTube to share pictures and video of the storm sweeping over Lake Erie. Among them were James Grimaldi, who sent up a drone to capture footage; two extreme athletes who posted footage of them surfing during the blizzard; and Alfonzo Cutaia, who posted a dramatic time-lapse video of the storm shot from his office to YouTube (Watson, 2014).

Cutaia, whose day job was that of an intellectual property lawyer, became infuriated at what he claimed as the misuse of his video, particularly by major cable networks. He brought first claims against CNN and CBC, alleging copyright infringement and when the first case was unsuccessful brought claims against CBS, Critical Mention Inc. and Mission Broadcasting in June 2016. He told First Draft News, a non-profit, which specialises in eyewitness media:

> People started reaching out to me in every way imaginable asking for permission to use it, but I quickly realised that many more were not asking permission. The most frustrating part was the lack of technical tools available to me for policing the unauthorised use of my video. So, I decided to use the tools of my trade instead.[1]
>
> (Dubberley, 2016:11)

This was not the first time copyright issues had been raised in reference to UGC and extreme events. Four years earlier, as the earthquake struck Haiti on 12 January 2010, Daniel Morel was busy taking photographs. His haunting pictures would appear on the world's front pages—but the newspapers didn't pay him, thanks to the decision he made a couple of hours later to post them via Twitpic.

Morel's images were reposted by a fellow Twitter user called Lisandro Suero, who claimed to have exclusive pictures of the quake; the newswire Agence France-Presse (AFP) picked them up and distributed them to their clients, while Getty Images disseminated them in the US. The pictures—showing Haitians in shock—were used by such prestigious media organisations as the *Washington Post*, CBS and ABC. AFP, Morel later complained, had sold his exclusive images for as little as \$45[2] (Estrin, 2013).[3]

Morel had a four-year fight in the courts to successfully prove that AFP and Getty had infringed his copyright by lifting his pictures from Twitter. But he was not an 'accidental journalist'—caught up in the disaster, and posting on social media because he had nowhere else to display his work. Morel was a professional photographer, a former Associated Press (AP) photojournalist, who had experience and knowledge about copyright in a long and distinguished career. He posted his pictures, expecting them to be picked up and credited as usual.

Yet what about the many others who are not media professionals but have produced powerful pictures or footage of disasters? If it took Morel, a photographer for more than two decades, nearly four years to make the courts, AFP and Getty acknowledge that he had copyright—and Cutaia, an intellectual property lawyer by trade some years to declare he was pleased with the outcome of his claim—citizen witnesses who upload pictures rather than sell them directly are even more vulnerable, despite supposed protection of the law.

Copyright law—along with laws of breach of confidence, passing off, trademarks and moral rights—is designed to prevent plagiarism and unfair exploitation of creative work, protecting creators of all kinds of work. But if it is applied too strictly, there is a danger that it may also inhibit the media's freedom to report and expose matters of public interest.

The first question to answer is whether copyright law applies to UGC—and the simple answer is that at present it is sometimes difficult to tell because of limited case law and different jurisdictions. As Julie Posetti of the Australian media giant, Fairfax Media, says: "The law is analogue and news is digital" (Dubberley, 2016:5). So, key questions that journalists who use other people's video, photographs or tweets should be asking themselves are as follows: what is the nature of the work; is it copyright protected; who was the author, and who is the rightful owner; what might exceptions/defences to copyright protection be; what possible problems with reuse and linking might there be (Cooper, 2017).

In the UK, for example, the *Copyright Designs and Patents Act* (1988) defines a work that can claim copyright as having two foci: the type of work it is and whether it can be said to possess some originality. Copyright is also protected under Article 1 Protocol 1 of the ECHR as a property right.

Content created in the aftermath of a disaster is often hurried, muddled and unedited—it is not carefully designed or created. But copyright often focuses on the *expression* of an idea rather than the idea itself, and does not concern itself with the quality or merit of a piece of work. So judgements from the European Court of Justice (ECJ) may favour creators of UGC. For example, in the case of *Painer v Standard VerlagsGmbH &Ors* [ECJ, 2011], Eva-Maria Painer, who had taken a portrait of Natascha Kampusch, the Austrian kidnap victim held in a cellar, and then found it reproduced without her consent in newspapers successfully sued media companies saying she was entitled to copyright. Pertinent to UGC creators the court ruled that because Painer had chosen how to frame the shot, the angle and the atmosphere, and then also selected which snapshot to use afterwards, even a standard portrait could be protected by copyright.

So filming a massive wave as it comes towards the shore, taking a picture of a collapsed house in the aftermath of an earthquake, and tweeting about your escape from a flood could all be said to involve effort and skill. Yet how much skill and effort is required—particularly with photographs that are taken rapidly, without long periods of preparation? Those caught in the middle of an earthquake or as a wave approaches do not have time to think about what kind of image they are producing.

The fact that much UGC is captured on mobile phone cameras is also problematic: these pictures are by nature ephemeral, can be deleted in seconds and have no negatives. To identify the owner of a photo's copyright what often needs to be established is whether a photo is original, whether it can be proved to be original and who the author is. This goes further than copyright and also embraces moral rights as well (that is, rights of attribution, the right to have it published anonymously or pseudonymously and the integrity of the work).

But the authorship, and provenance, of UGC photographs—both for creators claiming ownership or mainstream media looking for permissions—can be difficult to pin down, particularly during fast-moving news events.

The 2016 Brussels terror attacks were a good example of this. Two of the crucial videos—of the explosions at Zaventem airport and the aftermath—were attributed to journalists rather than those had actually shot them. Despite the social media agency Storyful tracking down the real creators, misattribution remained key. In these cases it seemed misunderstanding in a frantic situation had caused these errors—in the first case a journalist Anna Ahronheim had retweeted a video by Pinchas Kopferstein that had been shared in a Whatsapp group; in the second the journalist Steven Decraene had given clearance to media organisations after he'd been given permission to do so by its creator Aris Saja and the provenance had become confused (Cobben, 2016; Galvin, 2016). David Clinch of Storyful concluded that journalists were asking the wrong questions when seeking to establish the ownership of such content:

> Instead of asking 'can we use it?' journalists need to ask: where does this video come from, where were you when this happened? Do you have any other images to show that you were there?
>
> (quoted in Cobben, 2016)

Key here is the practice of 'scraping'. Embedding a tweet, picture or YouTube video onto a website is not a seen as a copyright infringement. In 2014, the European Court of Justice ruled this was the case following *BestWater International GmbH v Michael Mebes and Stefan Potsch (C-348/13)*. Mebes and Potsch had been embedding promotional videos made by BestWater circulated on YouTube into their own websites. The ECJ ruled that this was not infringing copyright because it was not transmitting the work to a new public or using a different method from the original communication. The court came to its conclusion also bearing in mind the CJEU ruling in *Svensson v Retriever Sverige AB* [EUECJ 2014][4] which was that once content has been made available to the public, it does not breach copyright to make it available by linking to it. But 'scraping'—removing the content from its original platform and then passing it off as your own on another or the same platform—however would be infringement.

Most discussions have unsurprisingly focused around visual media. But there is also the question about text as well. One of the media's current devices is to use tweets to add colour and immediacy to stories. But at only 140 characters long, journalists are often unclear whether tweets have the substance to be protected by copyright. Recent cases—such as *England And Wales Cricket Board Ltd & Anor v. Tixdaq Ltd & Anor* [2016], in which Arnold J found that the unauthorised reproduction and making available of clips of cricket matches lasting up to eight seconds amounted to an infringement of copyright—suggests that the media should be more careful.

Exxon Corp. v. Exxon Insurance Consultants International Ltd [1982] Ch. 119, found that one word was not copyrightable. But Edwards (2010) points out that when Peterson J used this quote from *University of London Press Ltd. v. University Tutorial Press Ltd. [1916] 2 Ch. 601* during the Exxon case it gave tweeters reason for hope: "There remains the rough practical test that what is worth copying is prima facie worth protecting" (1916:609–610 cited in Edwards, 2010).

The case of *Infopaq International v Danske Dagblades Forening* [EUECJ, 2009][5] could prove relevant for media organisations that extract from blogs or even tweets. Infopaq was a media digest business that was extracting short snippets (around 11 words) from newspaper articles in order to make searching and scanning easier for their clients—but was not paying copyright fees. The court ruled that while words themselves are not protected, the way a writer has arranged them is, even in such short extracts (see paras 45–50 in particular).[6] The extensive *Meltwater* litigation[7] went further. It suggested headlines can be copyrighted too, and the activities of a press-clipping service could not be seen as 'fair use'. In a related case, however, the Court of Justice of the European Union (CJEU) held that on-screen and cached copies of websites generated by users while browsing the internet do not infringe the copyright of publishers, and can be produced without authorisation or a licence from the right holder.[8]

Edwards debates in that case whether copying jokes and tweeting them breaches copyright (comedians put time and effort into crafting gags).[9] She asks the question: does copyright cover any tweet and, if so, what happens with retweeting. She concludes:

> Yet if copyright exists in tweets, such activity is *prima facie* copyright infringing. . . . Again, there is a strong argument that by writing in an unprotected, open to the public, Twitter account, you are granting an implied license to copy. . . . However the 'implied license' argument has been frequently repelled on the Net generally: it is now very well accepted that simply posting something on a website, like a photo, or a story, does not in any way grant permission to all and large to reproduce it. . . . Why should Twitter be any different?
>
> (Edwards, 2010)

Svensson v Retriever Sverige AB [2014] (EUECJ, 2014) complicates this, by asking whether copyright applies if the tweet is reproducing a link—and that link is not being introduced to a new public. Does a retweet create a deep link to the tweet, or reproduce the text? The answer will dictate whether a retweet is infringing EU copyright law.

One solution would be to stop the mechanism of the retweet (although I would argue that the retweet at least credits the original tweeter). But Edwards (2010) says by presumptively granting copyright to tweets, this would allow particular tweets to be easily suppressed from distribution on threat of legal action, something that might have serious and chilling consequences for freedom of speech, despite public interest defences.

Defences: Freedom of Expression and Public Interest

However, there are some elements of copyright law that mean the use of copyrighted work is permitted by the law. Copyright, although protected under Article 1 of the European Convention on Human Rights, cannot necessarily overrule Article 10, which deals with freedom of expression. This was tested in the *Ashby-Donald* case [2013][10] when the court found that copyright did not trump freedom of expression on that occasion, but in both that case and also in *Rocknroll v NGN Ltd* [EWHC 2013][11] it was clear that there could be instances where the reverse would be true.

The idea of public interest can also mean that copyrighted material is permitted to be used—as in *British Broadcasting Corporation (in the case of HM Advocate v Kimberley Mary Hainey) Petitioners* [2012] when the BBC persuaded a judge to release pictures of two-year-old Declain Hainey. The photographs, which had been shown to the jury, were not originally given to the media because the Crown Office argued that the copyright was owned by Kimberley Hainey, Declan's mother, who had taken the pictures (and who was also convicted of his murder, although the conviction was quashed on appeal).

The BBC argued that this was a high-profile case, which raised issues of serious public interest, that it had a public duty to report matters of public justice, and that the authorities' refusal to release the pictures was an unjustifiable interference with the media and public's right to freedom of expression under Article 10.

Lord Woolman ordered the release of six photographs that showed Declan Hainey alone. His justification was that Section 45 of the Copyright, Designs and Patents Act means copyright is not infringed when material is used for reporting parliamentary and judicial process. The BBC also referred to section 171(3) of the 1988 Act, which says that copyright can be trumped on the grounds of public interest, an argument the judge agreed with, citing *Ashdown v Telegraph Group Ltd* [EWHC 2001] and *Tarsasag A Szabadsagjogokert v Hungary* [ECHR 2011][12] as allowing greater press freedom and making sure the "law cannot allow arbitrary restrictions which may become a form of indirect censorship" (para 27, cited by Lord Woolman in para 37).

But with UGC, the photographer has not set out in advance to photograph a specific event or occurrence (natural disasters are by their very nature unpredictable). Does that mean that they have the same copyright protection? Singer and Turrell (2006) argue that it can, albeit using an example from the US. They cite the case of Abraham Zapruder, who set up his 8-millimetre movie camera to film President John F Kennedy passing by in his car—not intending, as it turned out, to film the president's assassination. The US courts ruled he had used sufficient skill and judgement in setting up his camera to be able to claim copyright, even though the subject matter had changed.

The media in the UK (and similar jurisdictions such as Australia, Canada and New Zealand) also have another defence for infringing copyright—what is known as: 'fair dealing'. Less flexible than the US concept of fair use, fair dealing covers use of copyrighted material for research, criticism and reviews, parodies but most pertinently for news reporting. Photographs however are exempted from this.

The sudden and unpredictable nature of natural disasters means that UGC gathered during them is likely to fall under the heading of 'current event', which gives the media a defence. Even UGC taken at the time of a disaster but used later may still be seen as current (for example, six months on from a disaster, a journalist investigating how aid has been spent in the year since uses some disaster footage). The cases of CCTV footage of Diana, Princess of Wales and Dodi Fayed in *Hyde Park Residence Ltd v Yelland* [EWCA 2000] and leaked notes of a meeting between Tony Blair and Paddy Ashdown in *Ashdown v Telegraph Group Ltd* [EWHC, 2001] were all arguably defined as current even though they emerged one and two years after the actual events.

Permission

But is it fair to (re)publish work from internet sites with the defence of reporting current events, if that material was not made widely available when first produced? It may seem self-evident that anything on an internet site is already published, and the CJEU's ruling in *Svensson v Retriever Sverige AB* [2014] (EUECJ, 2014) was that once content has been made available to the public, it does not breach copyright to make it available by linking to it. But it can be argued that the originator did not intend for it to be published more widely—the defence in the case of *HRH Prince of Wales v Associated Newspapers* [EWCA 2006] in which diaries of the Prince of Wales had been obtained by the *Mail on Sunday*—and which many of the UGC creators in the previous chapter were surprised at how far their unguarded comments or photographs had travelled.

From a self-regulatory point of view, good practice would be to regard some pictures as, in the BBC's words, "hidden in the open" (BBC, 2015)—intended to be shared no wider than family—and to bear in mind the intentions of the creator and the impact of reuse. The moral right of attribution—allowing the creator of a work to have their name attached (or, equally, not attached)—is also relevant here.

Many of the UGC creators that I spoke to were horrified at the thought of profiting financially from a disaster, but felt uneasy about the fact that their content had been used without their consent. This brings in the issue of not just legal protection but professional norms.

There is extensive legal literature on consent where the subject is often termed 'full and informed consent' and takes in misuse of information, breach of confidence, copyright and data protection law. Townend (2014) points out that consent has also been used as a defence in defamation and that it was touched on in the *Reynolds* privilege, which looked at "whether subjects had been approached for comment and the way in which their contribution was used" (2014:27).[13]

But consent as it applies to journalism practice goes beyond legal constraints and raises ethical problems, especially in the aftermath of disasters when an eyewitness/survivor/victim's ability to fully comprehend the consequences of what they are doing may be diminished. (The example of Jordi Mir, discussed in more detail later in this chapter, is a case in point—he talked about his decision to upload footage of the aftermath of the Charlie Hebdo attacks as a 'reflex' reaction). While obtaining consent is a basic ethical requirement for journalists, it is often implicitly rather than explicitly understood, and there is limited research into the subject. Muller (2013:37), in his investigation into the 2009 Australian 'Black Saturday' bushfires, puts forward the idea of the 'four abilities' model described by Grisso and Appelbaum (1998, cited in Muller, 2013). This measures people's capacity to consent by assessing their ability: 1. to make a choice; 2. understand the meaning of what is being proposed; 3. to appreciate the implications and consequences; and 4. when equipped with the necessary facts, to be able to arrive at a reasoned decision.

Muller talked to the survivors of the fires and found that while most had ability 1, not everyone had abilities 2 and 3, and ability 4 was almost entirely absent. On the basis of his findings, he argues that simple consent (where the journalist explains the probable use of the material) or informed consent (full details of the context, portrayal risks and benefits of being involved are outlined) are both impractical and simple consent, due to the survivors' state of mind, probably impossible (Muller, 2013:39–42). The result is that a kind of 'instinctual consent', where survivors gave consent without being able to fully appreciate the consequences but felt vindicated afterwards in their choice, prevailed. This in turn puts the onus on the journalist to recognise and respect the survivor's autonomy.

At this point some researchers may feel it is wrong for journalists to approach survivors because of the dangers; however, in an important news story such as a humanitarian crisis, it would be wrong to erase the voices of the victims and leave them silent. Such an approach would take us back to the 1980s and Michael Buerk at Korem, where no victims of the famine were interviewed during his seminal piece in 1984. What became clear from the interviews with UGC creators is that, while significant progress has been made in the way journalists approach the grief-stricken in real life (McLellan, 1999; Ewart, 2002; Townend, 2012; Newton and Duncan, 2012), the interaction in cyberspace limps behind, with consent often absent or perfunctorily obtained.

Privacy

> There's no doubt most teenagers don't think when they post to Facebook they don't think it is public . . . they see it as a public private space. It's like a conversation in the pub—it's in a public space, but not everyone and their uncle can eavesdrop.
>
> Kevin Marsh, BBC College of Journalism.
> (quoted in Whittle and Cooper, 2009:37)

What should the appropriate restrictions on journalists infringing the privacy of UGC creators be? This is a fraught area, which encompasses legal, ethical and professional norms.

When a crime, tragedy or natural disaster occurred in the past, a journalist would typically want to get their hands on a photo album or delve into a person's private life to discover a key detail to raise a story from the mundane to the compelling. This would commonly involve the journalistic practices known as the doorstep and the deathknock. These terms refer to a reporter physically waiting outside someone's house for a comment, or trying to persuade a grieving family to talk about the deceased.

Over the past decade, the growth of social networking sites has changed all this. Today Facebook and Twitter have been plundered for words and images. Any 'collect' [a family photograph from the past] in a newspaper these days is likely to have come from Facebook. As Paul Fievez, a former night picture editor wrote:

> Within seconds of a story breaking, news and picture desks are all assigning reporters, photographers and picture researchers to log-in to Facebook, Twitter, Linked-In, Friends Reunited. All of the other social networks and personal websites are Googled and scoured for pictures and information. If there is a live web-cam, pictures are grabbed and frequently published without any regard to copyright. Likewise, any images on the social sites and personal blogs or websites of anyone involved are all also grabbed before anyone has a chance to close the site down, and are then published, syndicated, used on television, re-published or broadcast repeatedly.
>
> (Fievez, 2011)

In the aftermath of the Manchester terror attacks in 2017, many survivors and families of those caught up in the event found themselves targeted by journalists online, most notably Dan Hett, whose brother Martyn Hett was killed, although the family did not have confirmation of this for several days. Hett found himself not just subject to the usual doorstepping but being pursued online—he estimated more than 50 journalists had contacted him this way before he had even had official confirmation of his brother's death.[14]

But veteran *Guardian* investigative reporter David Leigh echoes the view of most journalists—if you put something in the public domain then it is fair game. As he said in an interview in 2009:

> If you want to put stuff about yourself up in the public domain . . . and people can make what use of it they like. I don't see any point in complaining about it. If you don't want the information out there, don't put it out there.
>
> (Whittle and Cooper, 2009:35)

And as one newspaper news editor put it: "As a news editor I would expect my reporters to find pictures on Facebook—a reporter has let themselves down badly if they don't do the Facebook search" (Cooper, 2012:223).

But Leigh and the news editor do not take into account the possibility a third party has uploaded material about an individual without consent. For example, newspapers could end up using a picture uploaded not to your Facebook account, but a family member's.

Privacy and 'Contextual Integrity'

The ethical situation as Leigh and the news editor see it—if you put anything in the public domain, then face the consequences—is not quite as clear-cut as that. We are still grappling with the notion of what privacy is—and isn't—in relation to social networking sites. As the privacy theorist Helen Nissenbaum points out, from a sociological point of view the fundamental problem is a breakdown in what she calls "contextual integrity" (Nissenbaum, 2004:138).

In brief, she explains that privacy means different things in different situations, and that it is violated when people do not respect two types of contextual norms—those of appropriateness (what information may be shared) and those of flow and distribution (with whom the information is shared). So you share problems with your sex life with your friends, your good works for charity with your elderly aunt, your medical health with your doctor, your financial problems with your bank manager and your work-related goals with your boss. In many circumstances it would be highly inappropriate to share your sex life with your elderly aunt, or your medical problems with your bank manager.

The problem with sites like Facebook is that this can easily happen. There is what Grimmelmann (2009) calls a flattening of relationships—you are either in contact with someone or you are not; there are not the fine divisions in social relationships that there are in real life. So you may be 'friends' with your friends, your elderly aunt, your doctor and your boss on Facebook—and without sufficient privacy controls they may all learn things about you that during your usual real-life interactions they would not.

When material is taken from social networking sites by the media, then this transgression of contextual norms is taken even further. As Nissenbaum points out, in real life as opposed to virtual life there is nowhere that is not governed by at least some informational norms: we behave in a different way at work, at school or in the shopping mall—and because we do, we deliberately conceal and reveal different information depending on where we are.

She quotes the philosopher Ferdinand Schoeman as follows:

> [A] person can be active in the gay pride movement in San Francisco, but be private about her sexual preferences *vis-a-vis* her family and co-workers in Sacramento. A professor may be highly visible to other gays at the gay bar but discreet about sexual orientation at the university. Surely the streets and newspapers of San Francisco are public places as are the gay bars in the quiet university town. Does appearing in some public settings as a gay activist mean that the person concerned has waived her rights to civil inattention, to feeling violated if confronted in another setting?
>
> (Schoeman cited in Nissenbaum, 2004:140)

It's easy to see how that 'confrontation' in a different setting could take the form of exposure via a social networking site. Peterson cites the example of the *Daily Mail*, which, in 2007, published dozens of photos of drunken girls. The *Mail* wrote: "Drunkenly dancing on tables or collapsing in the street used to be a source of acute embarrassment for young women the morning after the night before. Today, they are more likely to boast about it—to the world, with pictures—on social networking sites". The photos, says Peterson, had been lifted from a Facebook group called '30 Reasons Girls Should Call It A Night'. He claims one of the students pictured, taken by surprise as she had not posted the photos herself, then found herself besieged by calls from overseas organisations offering her money for sexually explicit interviews (Peterson, 2010:11).

While none of the UGC creators interviewed for this book had had anything as serious as their sexuality[15] 'outed' or apparently private and embarrassing photographs splashed across the newspapers, many expressed unease about the way their use of social media during a humanitarian crisis had led to information spreading further than they had ever imagined and the discomfort they felt as a result.

Mass sharing and comment effectively created a new artefact and the original context was lost. In the most extreme cases, this can result in widespread vilification and even loss of a job and social status, as in the case of Justine Sacco, a PR executive for a media company. In 2013, Sacco tweeted a tasteless joke about AIDS to her 170 followers while flying to South Africa; it was picked up by an editor at Valleywag, Gawker's (now defunct) Silicon Valley gossip and news website. It got picked up, started trending, and the hashtag #HasJustineLandedYet was created. Sacco was publicly fired (Ronson, 2015).[16] Interviewee 65's experience of the widespread sharing of his 'Vine' video of the Oklahoma hurricane detailed in Chapter 7, and what he saw as criticism, left him stressed and guilty; he deleted his videos and temporarily left social media as a result.

The US v UK Experience: Content Acquisition and Content Dissemination

It's worth pausing briefly here to consider the difference between the US and UK approach—and how a (self) regulatory framework concerning privacy needs to acknowledge and perhaps learn from both approaches.

In the US, Calvert described the idea of the 'voyeurism value' in the First Amendment. What he meant was that freedom of expression was being used as a rationale to justify a post-modern, mediated visual world of television and the internet, so much so that the First Amendment

> increasingly safeguards, or is called upon to safeguard, our right to peer and to gaze into places from which we are typically forbidden, and to facilitate our ability to see and to hear the innermost details of others' lives without fear of legal repercussion.
>
> (Calvert, 1999 cited in Calvert, 2004:721)

Unease about how this can be interpreted in the US not only encompasses the serious (such as the autopsy pictures of NASCAR driver Dale Earnhardt, which resulted in the so-called Earnhardt Family Protection Act passed by the Florida Senate to ensure that pictures of the late driver were not published) to the salacious (the victory of wrestler Hulk Hogan after the gossip site Gawker published his sex tape in *Bollea v Gawker Media LLC et al* [2012] but potentially raises questions about privacy on social networking sites.

The situation in the UK is much more a question of balancing Article 8 and Article 10 of the Human Rights Act 1998 as opposed to the supremacy of the First Amendment. The differences between the two countries can be typified by two cases: the use of StreetView by (American giant) Google (in the UK) and the Taste, Ties and Time researchers who used Facebook to carry out a longitudinal study into undergraduates in an unidentified New England private college (Lewis et al, 2008).

When StreetView was introduced in 2007, 200 complaints were soon made to the pressure group Privacy International, which were passed to the Information Commissioners Office. A village in Cambridgeshire also prohibited entry to the Google StreetView car. Google responded by blurring facial pictures and the Information Commissioner ruled that, while there was a small likelihood of privacy invasion, StreetView should not be stopped. Although as the novelist and critic John Lanchester pointed out:

> It might be entertaining to look at the now-ubiquitous photo of the man coming out of the strip club, and you might argue that he deserves whatever he gets; but it's not necessarily without consequences for him, whoever he might be. As for the woman who had moved to a secret address to escape a violent partner, but was clearly identifiable on StreetView, let's hope her complaint was acted on in time.
>
> (Lanchester, 2009:20)

Yet it can be argued that StreetView was providing a public service (and was merely adding pictures to publicly available information, i.e. people's addresses via the electoral register—although, of course, it is possible to opt out of the open register). And it is not prohibited, except outside certain military bases, to take pictures of houses in the UK.

But the problem here is not just content acquisition, it is also content dissemination, and its resulting accessibility. StreetView's publication of pictures caused unease; Lanchester and others looked at how a breach of someone's privacy could do them harm, but there is also the dignity-based theory of privacy (Zimmer, 2010)—the idea of being able to control the flow of personal information, whether tangible harm takes place or not.

The case of the 'Tastes, Ties and Time' (T3) researchers encapsulates this. In September 2008 a group of researchers publicly released data collected from the Facebook accounts of an entire cohort of college students; the idea was to follow the students throughout their college career, gathering supplementary information such as where they lived on campus so that questions about the relationships between social networks online and offline could be posed.

Realising that there were privacy concerns inherent in the process, the researchers proclaimed that no one could be "connected to an identity". They removed student names from the dataset and set up 'terms and conditions' for other researchers to sign before they could use the dataset, merely identifying it as an "anonymous northeastern American university".

The university was quickly identified as Harvard. This meant it would be possible to identify the students too, even without their names; for example, the one self-identified Albanian in the cohort would not have been difficult to track down.

While it is clear there had been grave difficulties with the methodology, approval process and data release of this particular piece of research, the T3 experience could equally apply to the use of social networking by the mainstream media. Zimmer (2010), in his critique of the contextual privacy failings of the T3 cases, concludes that there needs to be a better understanding of privacy in social networking:

> Future researchers must . . . recog[nize] that just because personal information is made available in some fashion on a social network does not mean it is fair game for capture and release to all.
>
> (Zimmer, 2010:323)

While Zimmer is talking specifically about academic research here, it could again be equally applied to the media.

There is also an ethical question over the personal information in a social networking site—including political views and sexual interests—being transmitted to a much wider audience. Some of those I interviewed had been attacked on social media because their content and profile had reached more people after being used by mainstream media. The interplay between mainstream media and trolls (Phillips, 2011, 2013) appears to be a common part of this, particularly 'RIP trolling', where trolls attacked those who they saw as 'grief tourists' as opposed to those with 'authentic' grief.

Privacy and the Media—Codes and Laws

Is there any defence for those who find their images and text scattered over the media in the aftermath of a traumatic situation? It is worth looking at the UK as an example. The argument of media organisations (and the self-regulatory body IPSO and its predecessor the Press Complaints Commission has often agreed) was that if people put information or images in the public domain, even if they intend them only to be published for a small audience, then they are running the risk that the mainstream media will use them for a wider audience—particularly if they can argue a public interest.

In early cases concerning the PCC which closed in 2014 (*Baskerville v Daily Mail* (PCC 2011) and *Goble v People* (PCC, 2009) the complainants had never intended the information to go beyond a small number of followers or had been secured behind privacy settings. The PCC's successor, IPSO, took a similar view. In a recent case, *Strachan v MailOnline* (IPSO, 2018a), a man who sent a thoughtless tweet after coverage of traffic chaos because of an attempted suicide on the M5 found his tweet becoming the basis of a story in MailOnline. Shaun Strachan had tweeted "Push him the **** off or pull him the **** down" after reading tales of how a man had caused a seven-mile tailback on the motorway because of threatening to jump off a bridge. Strachan said he had deleted the tweet rapidly after sending it, but that the Mail had featured it, a picture of him from his Facebook profile and published his full name. IPSO concluded that even if he had deleted it later, he had put the comment into the public domain and thus did not have a reasonable expectation of privacy.

Twitter is generally acknowledged to be a public forum (unless a user 'protects' their tweets). Facebook however is seen as much more private because of the general assumption that you have to be 'friends' with someone to see their posts. Yet cases considered by IPSO suggest that it is still seen as the public domain, particularly if a photo or post goes viral. For example, in the wake of the London Bridge terror attacks of 2017, Katharine Armstrong shared a picture of her brother leaving the scene, still holding his unspilt pint of beer. The picture was generally celebrated as an example of British phlegm in the face of terror.

Armstrong complained about the use of the photo and comments in *Armstrong v thesun.co.uk* (IPSO, 2017), but IPSO ruled that she did not have a reasonable expectation of privacy, because she had not only shared it with her friends but 'tagged' them in it, making the post available to friends of friends, one of which was a journalist on another publication.

Similarly, IPSO recently ruled on a case *Collins vs thesun.co.uk* concerning the reality star Ferne McCann (IPSO, 2018b). On Christmas Day 2017, she had 'liked' an Instagram picture posted by Tommy Collins, the brother of her previous boyfriend Arthur who had been convicted of grievous bodily harm. This story was picked up by *The Sun's* website and MailOnline early in January because McCann had distanced herself from her former boyfriend.

But Tommy Collins complained to IPSO that he had put privacy settings on his Instagram account and thus it was a breach of his privacy. *The Sun* said that it had been public and only set to private afterwards. IPSO could not judge on when the post was public but ruled as Collins had 699 followers, not all of whom were known to him (and including the *Sun* journalist) there was no breach of privacy.

Even if someone has died, it is permitted to harvest information from their networking sites, according to the PCC's judgement in *Rundle v The Sunday*

Times (PCC, 2010). The paper took information from a MySpace page after an Oxford undergraduate, Toby Rundle, was found hanged, including text saying he'd like to meet God "just to ask him what I've done wrong". His mother Deborah complained that the page had had been created some considerable time earlier as a spoof and thus did not give an accurate impression of her son, but the PCC ruled: "One of the hazards of posting information online is that it can remain permanently and publicly accessible, and that a degree of control is lost once it is uploaded" (PCC, 2010).

Yet there are limits. Speaking at a conference organised by Lexis Nexis in 2009, William Gore, public affairs director at the PCC said:

> What these rulings show is that users of social networking sites have a reasonable expectation that 'private' information they post online will be unlikely to find its way into the mainstream unless they, and/or the information they have published, are newsworthy. The ruling in the Mullan case[17] also sets out a clear warning to newspapers that it is not acceptable to trawl the net for information about anyone and everyone and then to publish material that is private and should remain so. Fundamentally, journalists should ask themselves the same questions as they should when deciding whether to publish a photograph taken by an old-fashioned pap. Where has the information come from? How has it been obtained? What exactly does it show?
>
> (Gore, 2009)

The law has been further strengthened concerning the privacy of children since then. The General Data Protection Regulation makes clear that consent for data collected about under-16s must be given by the child's parent or custodian—and the regulation states that data controllers must be able to prove this consent. Meanwhile, *Weller v Associated Newspapers Ltd* [EWHC,2014] in which the children of Paul Weller were photographed when on a private shopping trip, indicated the heightened privacy rights afforded to children upheld in *Murray v Express Newspapers plc* [EWHC,2008] despite the fact that the singer Paul Weller's wife had shared pictures of her children on social media.

The law of England did not specifically recognise a right to privacy until 2 October 2000 when the *Human Rights Act 1998* came into effect. Initially Article 8 seemed to take precedence, particularly following *von Hannover v Germany* [ECHR, 2004] (in which the ECHR ruled respect for the private life of Princess Caroline of Monaco had been breached by photographs of her shopping or on holiday in public places), *McKennitt v Ash* [EWHC, 2005] and *CC v AB* [EWHC, 2006] when a prominent sportsman secured an injunction after a betrayed husband wanted to publicise the sportsman's adulterous relationship with his wife.

But subsequent cases have seen courts emphasise Article 10 (freedom of expression)—those of *Axel Springer AG v Germany* [ECHR, 2012a] and

Von Hannover v Germany no 2 [ECHR, 2012b]. In the first, the European Court of Human Rights ruled that Article 10 was violated after the German newspaper *Bild* was prevented from revealing the arrest and conviction of a well-known actor for drugs possession. In the second, the magazine *Frau im Spiegel* had taken pictures of Princess Caroline on a skiing holiday while her father was very ill; she claimed this was a breach of Article 8, but the court ruled that the photographs contributed to a debate in the public interest.

But such cases tend to focus on public figures. What about ordinary members of the public? To take one possible scenario: if someone films the survivor of an earthquake, which is then used by the BBC and viewed by millions, has privacy been breached?

The PCC noted in its adjudication of *Kirkland v Wiltshire Gazette and Herald (2008)* "Rare and large-scale events such as terrorist attacks and natural disasters involve a degree of public interest so great that it may be proportionate and appropriate to show images of their aftermath without the consent of those involved" (PCC, 2008). Certainly the case of Armstrong v *thesun.co.uk* suggests this is the case.

But *Peck v UK* [ECHR, 2003] suggests that those who use UGC of dramatic events without seeking the permission of those they feature may lay themselves open to legal action. This case concerned Geoffrey Peck who, while depressed, had attempted to die by suicide in public—but was saved thanks to being spotted via CCTV installed by the local council. To show how useful the council's CCTV could be, the council later released photographs and short clips to local newspapers and TV.

The ECHR found that releasing these images without Peck's consent, and without masking his identity sufficiently infringed his right to privacy under Article 8.

This has however been seen as a problematic case. First it could be argued that natural disasters have a public dimension that suicides usually do not. Second, Robertson and Nicol argue that Peck "*created* a public event by his actions" (2007:275) [their emphasis]. The reality is, though, that courts would have to consider Peck's case in weighing up any action taken by for example, a victim of a humanitarian crisis, broadcast by a media organisation.

Yet a surprising conclusion by the Canadian Broadcast Standards Council (CBSC, 2012) also supports the use of content during what we might think of as a moment where people would be granted privacy: when seriously injured and close to death. The CBSC ruled on the decision by broadcaster CITY-TV to play audio of a police officer who was pinned under a mini-van and managed to radio for help (he later succumbed to his injuries in hospital). Another station, CTV Toronto, did not play the audio but displayed a written transcript on screen. The council ruled that, while there should have been more warning given, the stations did not breach the code of ethics either on privacy or decency—because the original police radio transmission could have been heard by the public as it was not encrypted in any way, and because it showed the professionalism and dangers faced by law officers in carrying out their duty.

One of the adjudicators, D Braun, however, dissented. He made the points that it was not up to television stations to broadcast the last words of a dying man, and even though police transmissions can be intercepted with scanners, it does not mean that police forces have given tacit approval to broadcast the content; consent should have been obtained beforehand.

This argument came into sharp focus again in 2015 after the use of Jordi Mir's footage in the aftermath of the attacks on the Charlie Hebdo offices in Paris. The shocking and brutal attacks resulted in an explosion of photographic and video content on social media, much of which was picked up by the mainstream media. Mir's, however, stood out because he filmed the attackers Cherif and Said Kouachi in the act of killing a prone police officer, later identified as Ahmed Merabet.

Mir later told AP that he had uploaded his film onto Facebook in a "stupid, reflex" action; he felt that he needed to talk to someone about what he had seen because he was alone in his flat, but that he took it down within 15 minutes (Satter, 2015). By then, however, he had lost control of the film: someone had uploaded it to YouTube and it was widely used by the media. A still from the clip made the front page of *Le Figaro*. The family of Merabet were traumatised by the continual reuse of the footage. Merabet's brother attacked journalists, saying:

> How dare you take that video and broadcast it? I heard his voice. I recognized him. I saw him get slaughtered and I hear him get slaughtered every day.
>
> (Alexander, 2015)

Mir turned down offers of payment, while authorising some media organisations to use the film as long as they cut the moment of death; some, he said, ran it without permission (Satter, 2015; Sargent, 2015).

Media Responses—Guidelines

The need for media responses to go beyond the legal and deal with the ethical issues is one that is becoming more discussed. Some media organisations are formulating guidelines to deal with people creating content during traumatic stories sensitively (the DART Centre led the way with much of this). In 2011, the media journalism site Journalism.co.uk published a checklist that journalists should go through before republishing content from social media (Marshall, 2011). It referred journalists to the BBC guidelines but also the [then] PCC Editors' Code and the Reuters Handbook of Journalism.

Complicating this, however, is the reality that codes are often interpreted very differently in practice according to institutional norms—not just different practice between press and broadcasters but also between quality and tabloid and public service and commercial. In comparison, the BBC, as a public service broadcaster, reflects in its guidelines on how material from social networking sites should be used. It comments:

> Whilst some in the media might argue that, once an individual has begun a declarative lifestyle . . . they cannot expect to be able to set limits on that, people making content for the BBC should ask themselves whether a door that is only ajar can justifiably be pushed further open by the media. . . . And it should be considered that the use of social media content by the BBC often brings that content to a much wider public than a personal website or social media page that would only be found with very specific search criteria.
>
> (BBC, 2010)

Graphic Imagery

> "What is that?"
> "Fuckkkk!!!
> "That's a body? Wow, just a hat and meat left".
> "Jesus"
> . . .
> "OMG I can't ever unsee that"
>
> (Comments left on Derek Van Pelt's Periscope stream of the 2015 Bangkok blast, collated by Brown, 2015)

With the 2004 tsunami, the first changes in how editors would respond to large-scale events in future became clear. As one Australian picture editor put it: "Photography played a crucial role in explaining the scale and impact of what had happened" (Meares, 2005).

With the tsunami's victims including many Westerners among them, mainstream media agonised over what it was possible to show. They made their decisions public in many cases, via columns, soliciting reader responses, while journalistic think tanks such as DART reissued lists of resources to deal with the sensitive situation. A year earlier, in the aftermath of the Oklahoma City bombing, DART had suggested to editors, when considering what images to use, to: "ask [themselves] whether the images are pertinent or will do unnecessary harm to certain members of your readership or broadcast audience (DART, 2003:3). In the UK, newspapers were guided by the PCC code and subsequently (for those who have signed up to it) the IPSO code, which said that "In cases involving personal grief or shock, enquiries and approaches must be made with sympathy and discretion and publication handled sensitively" (IPSO, 2016).

Section 8 of the Ofcom Broadcasting Code is more comprehensive, reflecting the more invasive and powerful nature of sound and visual images. It states:

> 8.16 Broadcasters should not take or broadcast footage or audio of people caught up in emergencies, victims of accidents or those suffering a personal tragedy, even in a public place, where that results in an infringement of privacy, unless it is warranted or the people concerned have given consent.
>
> 8.17 People in a state of distress should not be put under pressure to take part in a programme or provide interviews, unless it is warranted.
>
> 8.18 Broadcasters should take care not to reveal the identity of a person who has died or of victims of accidents or violent crimes, unless and until it is clear that the next of kin have been informed of the event or unless it is warranted.
>
> 8.19 Broadcasters should try to reduce the potential distress to victims and/or relatives when making or broadcasting programmes intended to examine past events that involve trauma to individuals (including crime) unless it is warranted to do otherwise. This applies to dramatic reconstructions and factual dramas, as well as factual programmes.
>
> In particular, so far as is reasonably practicable, surviving victims and/or the immediate families of those whose experience is to feature in a programme, should be informed of the plans for the programme and its intended broadcast, even if the events or material to be broadcast have been in the public domain in the past.
>
> (Ofcom, 2017)

The tension between the newsworthiness of events and the audience's sensibilities came to the fore during a terrorist event rather than a natural disaster. The BBC found itself in breach of Ofcom's code over its coverage of the 7/7 bombings. While the presenter did give a warning about the upsetting nature of what was to follow, viewers saw a victim being given heart massage at a London hospital. In its rush to get the pictures to air, the BBC later admitted it had not seen all the footage; Ofcom censured the corporation for not giving a strong enough warning (Ofcom, 2005).

Yet during the tsunami, while UK media were careful about what they showed of (Western) casualties, this reticence became increasing irrelevant as images—many more graphic than the media were showing—were posted to multiple non-news sites within hours of being taken, and were viewed by millions, "leading the news organisations of old media to realise the coverage was only partly in their control" (Zelizer, 2010:247). For example, one website—waveofdestruction.org—put up by Australian blogger Geoffrey Huntley had more than 682,000 unique visitors in just four days (Cooper, 2007).

The 2004 tsunami (what Kennicott (2010) called a "seemingly transparent window on misery") was far from isolated in this regard. Other key events include the death of Saddam Hussein, in which journalism's efforts at "gatekeeping are now almost entirely irrelevant. The public will find exactly as much of the death of Hussein as it wants and people will watch for as long as it holds any novelty or fascination" (Kennicott, 2006).

Yet Kennicott's piece was filed before mobile phone footage of Saddam's death emerged. It told a very different story from the official version, which had been sombre and appeared to show Saddam going to his death quietly. This time with sound, it revealed the taunts of the executioners, shouting at Saddam that he should go to hell and chanting the name of the Shia militant Moqtada al-Sadr. Mainstream media were caught on the hop and had to revise their earlier accounts Saddam's death.

The result is that editors' usual approach—limited by what readers/viewers can bear to see over a mythic breakfast table, in line with the Ofcom/IPSO standards—is being challenged. As Zelizer puts it:

> Journalists often avoid depicting what they think is most problematic, but as recent events involving citizen journalists show, non-journalists may have no such reticence. When people other than journalists can exploit the porousness of images of impending death and distribute them at will, decisions about what to show can be taken in journalism's name but without journalism's sanction to varying effect.
>
> (Zelizer, 2010:266)

The fact that the creators of this content are not 'professional' journalists means they are not subject to the same considerations of privacy and taste that, for example, the BBC demands. The BBC's Editorial Guidelines make clear that images and words must be chosen with care when covering stories that report suffering and distress. They state:

> We will always need to consider carefully the editorial justification for portraying graphic material of human suffering and distress. . . . It is always important to respect the privacy and dignity of the dead. We should never show them gratuitously. We should also avoid the gratuitous use of close ups of faces and serious injuries or other violent material.
>
> (BBC Editorial Guidelines, 2018)

Legally speaking, although one could argue that more gruesome pictures could be subject to Article 8 of the *Human Rights Act* (namely, that victims of disasters should be afforded privacy and respect), laws in the UK relating to taste, decency and obscenity tend to concentrate on sexual (im)morality. It is difficult to imagine that a case could be brought on the grounds that it would deprave or corrupt (via the Obscene Publications Act) against a news organisation that broadcast gruesome images as part of a news story; the only recourse to complaint would be under the regulatory codes.

The concern is that Kennicott's "window on misery" is being rendered ever more transparent by the pressures put on the mainstream media by a glut of UGC. The (controversial) decision of most mainstream newspapers to put a picture taken from mobile phone stills of a dead Muammar Gaddafi on their front pages, shows the distance that newspapers have come since 2005/06: there has been a shift in public sentiment about what news images are for and "a diminishment of journalists' formerly exclusive and authoritative call on the images they show, and shifts toward a larger and more vocal public role in evaluating pictures" (Zelizer, 2010:52).

Further footage of Gaddafi being beaten, assaulted and finally killed in Sirte was used extensively in the British media, despite the fact that many of the images were extremely gruesome. The columnist Suzanne Moore complained "A snuff fest is upon us via every possible medium" (Moore, 2011). But the *Guardian*'s Jonathan Jones claimed that those in the West who complained about the proliferation of such images were guilty of a hypocritical and self-deceiving reaction: that the reality of war had to be shown rather than hidden away (Jones, 2011).

The result is that news organisations are making constant ethical statements about what they are willing to publish or not publish. If they use UGC which shows imagery of disaster or death, they may to choose to frame it by cropping, using black and white instead of colour, using it only on certain parts of a bulletin, paper or website or prefacing this with 'disturbing content', they are implicitly and explicitly referring to their editorial judgements on what is 'acceptable' or not.

Conclusion

User-generated content is now a firmly established part of disaster reporting in the mainstream media. It has been rapidly absorbed into the journalists' lexicon. The idea of reporting an earthquake, a tsunami, a flood or a cholera outbreak and not using footage filmed or tweets written by amateurs seems unthinkable for most media professionals, even if they perceive it as just another source to be mined as opposed to a tilting of the journalistic field. The result has been a smash-and-grab attitude to the public's words and images, often while paying lip service to consent, privacy and taste.

This chapter aimed to point out some of the unresolved issues that this plundering has caused in order to examine more closely in following chapters how journalists saw the use of UGC. While copyright has been widely accepted in the use of images, with text it continues to be contentious and the application of copyright law to user-generated content remains sketchy. The basics of consent still appears to be a fraught area, with confusion over whether republishing requires additional permission. With successive cases, privacy appears to have moved on from the argument that if you put something online it is fair game. But the complications of other people's access to an individual's online pictures, coupled with the idea that a new artefact may be created once an image or post has been shared and commented on, that is completely devoid of context, means that much still rests on normative and deontological journalistic approaches.

The growth of platforms such as Periscope or Facebook Live suggest that our chances of being confronted by graphic content is higher; while a generation ago, a photographer tended to capture the aftermath of a disaster, now there is a reflex reaction in those witnessing it to switch on their phones and start recording: Tony Kemp, the first medic at the scene of the Shoreham aircrash in August 2015, said those doing so were in 'shock' but police warned against posting such graphic footage online (Merrill, 2015). Suicides, murders, terrorist attacks have all become Facebook Live 'events'. In 2017, four people in Chicago were charged with hate crimes, kidnapping and battery after a Facebook Live video showed them beating a disabled man (Levin and Jamieson, 2017), while in 2016 a double murder of police employees in Paris was broadcast (Toor, 2016).

Finally, the ethical debate in this chapter may have focused on new technologies but the questions remain the same as they have ever been. Journalists should respect privacy, credit other people for their own work, think carefully before releasing violent content, rather than allowing it to circulate.

Notes

1. In May 2018, Cutaia confirmed with regard to the status of his copyright matters, all had been concluded.
2. Morel's quote in Estrin (2013); he does not specify whether Morel is talking about exclusive rights here
3. Morel successfully sued AFP and Getty Images—see *Agence France Press v Morel* [[SDNY, 2013]]. He won $1.22m after a jury ruled that the organisations had wilfully infringed his copyright of eight pictures
4. Articles written by several Swedish journalists were published on the website of the Göteborgs-Posten. However Retriever Sverige, a Swedish company, operates a website that provides its clients with hyperlinks to articles published on other websites including Göteborgs-Posten, but without asking the journalists' permission to do so. The Svea Hovrätt (Court of Appeal in Sweden) brought the matter to the CJEU to see whether this constitutes an 'act of communication' under EU Law—which would mean that it would not be possible to make the hyperlinks available without the copyright holders' permission. The CJEU established that it did—but that the communication must be directed at a new public. In this case there was no new public as the Göteborgs-Posten website was freely accessible.
5. Case C-5/08 [2009] ECR I-6569, paragraph 37
6. Xalabarder notes that the ECJ conceded that "words . . . considered in isolation are not as such an intellectual creation of the author who employs them. It is only through the choice, sequence and combination of those words that the author may express his creativity in an original manner and achieve a result which is an intellectual creation" (Infopaq, 2009:para 45) (Xalabarder, 2012).
7. *Newspaper Licensing Ltd & Others v Meltwater Holding & Others* [2010] EWHC 3099. Meltwater, an online media monitoring service, provided its subscribers with a report containing the headline, an extract of the article and a hyperlink to the publisher's website. The UK court concluded that the headline

could be subject to copyright, and that the press-clipping service could not qualify as a fair dealing since the websites of the newspaper publishers expressly stated that their contents could not be used for any commercial purposes.

8. The Newspaper Licensing Agency (NLA) had argued Meltwater was infringing copyright because such copyright subsisted in the cached copies of the work that saved onto users' computers while they browsed the web. The High Court and Court of Appeal held that the users of the service did require a licence from the NLA on the grounds that the temporary copying defence under Article 5(1) Directive 2001/29/EC (the Directive) had not been made out. The case then went to the Supreme Court of Justice, which referred it to the CJEU. The CJEU, however, decided that the on-screen and cached copies were temporary within the meaning of the Directive.
9. In fact in July 2015, Twitter responded to a Digital Millennium Copyright Act (DMCA) take-down notice by removing tweets containing a joke on the basis of copyright infringement. A writer called Olga Lexell had posted a tweet which was then reposted by several other Twitter users without crediting her and she appealed to Twitter. Twitter proceeded to remove the offending tweets and they were replaced with a message stating "*this tweet has been withheld in response to a report from the copyright holder*". See Parkinson (2015).
10. ECtHR (5th section), 10 January 2013, case of *Ashby Donald and others v. France*, Appl. nr. 36769/08. This concerned fashion photographers who had taken pictures of fashion shows and published them on a website without consent of the fashion houses.
11. Ned RocknRoll, the partner of actress Kate Winslet, sought an injunction to stop pictures of him at a fancy dress party, which had been shared on a friend's Facebook page. *The Sun* was tipped off about these pictures and sought to publish them. RocknRoll sought the injunction on two grounds: privacy, and also because he had obtained the copyright for the photos.
12. In this case, the ECHR interpreted Article 10 and implicitly recognised the right of access to official documents. The applicant was a civil liberties organisation, which asked Hungary's Constitutional Court to disclose a parliamentarian's complaint questioning the legality of new criminal legislation concerning drug-related offences. The ECHR pointed out that NGOs, like the media, can have a 'watchdog' function and the state should allow them access to official documents. It concluded that "the interference with the applicant's rights was a violation of Article 10 of the Convention" (Voorhoof, 2009).
13. This has now been replaced by the new public interest defence in the Defamation Act 2013.
14. See his tweet https://twitter.com/danhett/status/866960257150529536?lang=en
15. This would potentially fall under the tort of misuse of private information, which focuses on "the protection of human autonomy and dignity—the right to control the dissemination of information about one's private life and the right to the esteem and respect of other people" *Campbell v MGN [2004] 2 AC 457.*
16. Sacco was rehired by the company in January 2018 (Wagner, 2018). https://www.recode.net/2018/1/19/16911074/justine-sacco-iac-match-group-return-tweet.
17. Mullan, Weir, Campbell (PCC, 2009). The *Sunday Express* had trawled Facebook for comments and pictures made by the Dunblane survivors as they turned 18, and wrote a piece criticising their drinking habits.

References

Alexander, H. 2015, 11 January-last update, *Funeral for French policeman Ahmed Merabet held in Paris* [Homepage of Telegraph], [Online]. Available: www.telegraph.co.uk/news/worldnews/europe/france/11338404/Funeral-for-French-policeman-Ahmed-Merabet-held-in-Paris.html [2015, 2 November].

BBC. 2010, October-last update, *Social media and third party websites: Use of pictures* [Homepage of BBC], [Online]. Available: www.bbc.co.uk/editorialguidelines/page/guidance-social-media-pictures#editorial-guidelines-issues [2015, 8 January].

BBC. 2015, 26 March-last update, *Social media guidance for staff* [Homepage of BBC], [Online]. Available: http://news.bbc.co.uk/1/shared/bsp/hi/pdfs/26_03_15_bbc_news_group_social_media_guidance.pdf [2016, 1 April].

BBC. 2018, *Editorial guidelines: Reporting death, suffering, distress* [Homepage of BBC], [Online]. Available: www.bbc.co.uk/editorialguidelines/guidelines/privacy/reporting-death [2018, 8 May].

BestWater *International GmbH v Michael Mebes, Stefan Potsch* 2014, C343/13 edn.

British *Broadcasting Corporation (in the case of HM Advocate v Kimberley Mary Hainey) Petitioners*—[2012] HCJDV 10 2012, High Court of Justiciary.

Bollea v. Gawker Media LLC, 2012 CA01244.

Brown, P. 2015, 25 August-last update, *OMG I can't ever unsee that: What happened when the live aftermath of the Bangkok blast was streamed on Periscope* [Homepage of Eyewitness Media Hub], [Online]. Available: https://medium.com/1st-draft/omg-i-can-t-ever-unsee-that-what-happened-when-the-aftermath-of-the-bangkok-bomb-blast-was-7a3f39ee2b0 [2016, 7 January].

Calvert, C. 2004, "Revisiting the Voyeurism value in the first amendment—from the sexually sordid to the details of death", *Seattle University Law Review*, vol. 27, pp. 721–749.

CBSC. 2012, 7 March-last update, *CBSC decision 10/11–2185 & -2186* [Homepage of CBSC], [Online]. Available: www.cbsc.ca/ciii-tv-global-toronto-chch-tv-rebroadcasts-of-a-police-radio-transmission/ [2018, 8 May].

Cobben, I. 2016, 24 March-last update, *Mass misattribution of viral Brussels video* [Homepage of World News Publishing Focus], [Online]. Available: http://blog.wan-ifra.org/2016/03/24/mass-misattribution-of-viral-brussels-video [2016, 8 April].

Cooper, G. 2007, *Anyone here survived a wave, speak English and got a mobile? Aid agencies, the media and reporting disasters since the tsunami*, Nuffield College, Oxford.

Cooper, G. 2012, "Facing up to the ethical uses of Facebook", in *The phone hacking scandal: Journalism on trial*, eds. R. Keeble & J. Mair, Abramis, Bury St. Edmunds, pp. 353–230.

Cooper, G. 2017, "Rights and responsibilities when using user-generated content to report crisis events", in *The Routledge companion to media and human rights*, eds. H. Tumber & S. Waisbord, Routledge, London, pp. 257–268.

DART Center for Journalism and Trauma. 2003, *Tragedies and journalists* [Homepage of DART Center], [Online]. Available: https://dartcenter.org/sites/default/files/en_tnj_0.pdf [2018, 8 May].

Dubberley, S. 2016, "A journalist's guide to copyright law and eyewitness media", *First Draft News*. Available: https://firstdraftnews.org/wp-content/uploads/2016/11/1stdraft_copyright.pdf [2018, 2 July].

ECHR. 2003, *Peck v the United Kingdom*. ECHR 44. Available: www.bailii.org/eu/cases/ECHR/2003/44.html [2017, 5 March].

ECHR. 2004. *Von Hannover v Germany*. EMLR 21. Available: www.bailii.org/eu/cases/ECHR/2012/228.html [2017, 5 March].

ECHR. 2011. *Társaság A Szabadságjogokért v Hungary*. 53 EHRR 3, [2009] ECHR 618. Available: www.bailii.org/eu/cases/ECHR/2009/618.html [2017, 5 March].

ECHR. 2012a, Axel *Springer AG v Germany* [2012] EMLR 15. Available: www.bailii.org/eu/cases/ECHR/2004/294.html [2017, 5 March].

ECHR. 2012b. *Von Hannover v. Germany (no. 2)* [2012] ECHR 228. Available: www.bailii.org/eu/cases/ECHR/2012/227.html [2017, 5 March].

ECJ. 2011, *Painer v Standard Verlags GmbH* [2011] EUECJ C-145/10. Available: www.bailii.org/eu/cases/EUECJ/2011/C14510.html [2017,5 March].

ECtHR (5th Section) 2013. *Ashby Donald and others v. France*, Appl. nr. 36769/08. Available: http://hudoc.echr.coe.int/eng?i=001-115845 [2018, 28 August].

Edwards, L. 2010, 22 July-last update, *We are not amused: Jokes, Twitter and copyright* [Homepage of Lilian Edwards blog], [Online]. Available: http://blogscript.blogspot.co.uk/2010/07/we-are-not-amused-jokes-twitter-and.html. [2012, 7 May].

England *and Wales Cricket Board Ltd and Another v Tixdaq Ltd and Another*. 2016 EWHC 575 (Ch) 2016, Chancery Division.

Estrin, J. 2013, 23 November-last update, *Haitian photographer wins major US copyright victory* [Homepage of New York Times], [Online]. Available: http://lens.blogs.nytimes.com/2013/11/23/haitian-photographer-wins-major-u-s-copyright-victory/?_r=0 [2015, 8 January].

EUECJ. 2009, *Infopaq International A/S v Danske Dagblades Forening* [2009] EUECJ C-5/08 (16 July 2009). Available: www.bailii.org/eu/cases/EUECJ/2009/C508.html [2018, 30 August].

EUECJ. 2014a, *Svensson and others v Retriever Sverige AB* [2014] 3 CMLR 4. Available: www.bailii.org/eu/cases/EUECJ/2014/C46612.html [2017, 5 March].

Ewart, J. 2002, *Prudence not prurience: A framework for journalists reporting disasters*, Proceedings of the 23rd Annual Conference of the Australia and New Zealand Communication Association (ANZCA 2002).

EWCA. 2000 *Hyde Park Residence Ltd v Yelland & Ors* [2000] EWCA Civ 37. Available: http://www.bailii.org/ew/cases/EWCA/Civ/2000/37.html [2018, 30 August].

EWHC. 2001, *Ashdown v Telegraph Group Ltd* [2001] EWHC Ch 28. Available: www.bailii.org/ew/cases/EWHC/Ch/2001/28.html [2018, 30 August].

EWCA. 2006 *Associated Newspapers Ltd v HRH Prince of Wales* [2006] EWCA Civ 1776. Available: http://www.bailii.org/ew/cases/EWCA/Civ/2006/1776.html [2018, 30 August].

EWHC. 2005 *McKennitt v Ash* [2005] EWHC 3003 (QB). Available: http://www.bailii.org/ew/cases/EWHC/QB/2005/3003.html [2017, 5 March]

EWHC. 2006 CC *v AB* [2006] EWHC 3083 (QB). Available: http://www.bailii.org/ew/cases/EWHC/QB/2006/3083.html [2017, 5 March].

EWHC. 2007 Murray v Express Newspapers plc& Anor [2007] EWHC 1908 (Ch). Available: http://www.bailii.org/ew/cases/EWHC/Ch/2007/1908.html [2018, 28 August].

EWHC. 2010 *The Newspaper Licensing Agency Ltd & Ors v Meltwater Holding BV & Ors* [2010] EWHC 3099 (Ch). Available: http://www.bailii.org/ew/cases/EWHC/Ch/2010/3099.html [2018, 28 August].

EWHC. 2013 *Rocknroll v News Group Newspapers Ltd* [2013] EWHC 24 (Ch) Available: www.bailii.org/ew/cases/EWHC/Ch/2013/24.html [2018, 30 August].

EWHC. 2014 Weller v Associated Newspapers [2014] EWHC 1163 (QB). Available: http://www.5rb.com/case/weller-ors-v-associated-newspapers-limited/ [2018, 28 August].

Exxon Corporation And Others v. Exxon Insurance Consultants International Ltd. [1982] Ch. 119 1981 [Court of Appeal].

Fievez, P. 2011, 11 November-last update, *Borrowed photos (2)* [Homepage of Gentlemenranters.com], [Online]. Available: http://gentlemenranters.com/page_325.html#pf220 [2011, 30 November].

Galvin, J. 2016, 31 March-last update, *Brussels and the perils of rip and ship journalism* [Homepage of Storyful], [Online]. Available: https://storyful.com/blog/2016/03/31/brussels-and-the-perils-of-rip-and-ship-journalism/ [2018, 30 April].

Glocer, T. 2006, 11 October-last update, *We media speech* [Homepage of Tom Glocer], [Online]. Available: http://tomglocer.com/blogs/sample_weblog/archive/2006/10/11/98.aspx [2012, 20 May].

Gore, W. 2009, November-last update, *Privacy and the PCC—adapting to changing circumstances* [Homepage of Press Complaints Commission], [Online]. Available: www.pcc.org.uk/news/index.html?article=NjA3OQ== [2012, 20 May].

Grimmelmann, J. 2009, "*Saving Facebook* NYLS legal studies research paper no. 08/09–7", *Iowa Law Review*, vol. 94, p. 1137.

Grisso, T. & Appelbaum, P. 1998, *Assessing competence to consent to treatment*, Oxford University Press, Oxford.

Hermida, A. & Thurman, N. 2008, "A clash of cultures", *Journalism Practice*, vol. 2, no. 3, pp. 343–356.

IPSO. 2016, 1 January-last update, *Editors' code of practice* [Homepage of IPSO], [Online]. Available: www.editorscode.org.uk/the_code.php [2016, 15 April].

IPSO. 2017, 1 September-last update, *Armstrong v thesun.co.uk 12990–17* [Homepage of IPSO], [Online]. Available: www.ipso.co.uk/rulings-and-resolution-statements/ruling/?id=12990-17 [2018, 8 May].

IPSO. 2018a, 29 January-last update, *Strachan v MailOnline* [Homepage of IPSO], [Online]. Available: www.ipso.co.uk/rulings-and-resolution-statements/ruling/?id=20529-17 [2018, 8 May].

IPSO. 2018b, 9 March-last update, *Collins v thesun.co.uk 20838–17* [Homepage of IPSO], [Online]. Available: www.ipso.co.uk/rulings-and-resolution-statements/ruling/?id=20838-17 [2018, 8 May].

Jones, J. 2011, 25 October-last update, *The west wrings its hands over dead Gaddaffi pictures* [Homepage of The Guardian], [Online]. Available: www.guardian.co.uk/commentisfree/2011/oct/25/dead-gaddafi-photos-war-hell [2011, 11 December].

Kennicott, P. 2006, 31 December, "For Saddam's page in history, a final link on YouTube", *Washington Post*, Washington DC, p. D01.

Kennicott, P. 2010, 16 January, "Graphic visuals in Haiti coverage break from tradition", *The Washington Post*, Washington DC, p. C10.

Lanchester, J. 2009, "Short cuts", *London Review of Books*, vol. 31, no. 7, p. 20.

Levinson, S. & Jamieson, A. 2017, 5 January-last update, *Four suspects charged with hate crimes over beating in Facebook live video* [Homepage of The Guardian], [Online]. Available: www.theguardian.com/us-news/2017/jan/05/facebook-live-beating-anti-donald-trump [2018, 27 April].

Lewis, K., Kaufman, J., Gonzalez, M., Wimmer, A. & Christakis, N. 2008, "Taste, ties and time: A new social network dataset using Facebook.com", *Social Networks*, vol. 30, no. 4, pp. 330–342.

Marshall, S. 2011, 17 August-last update, *How to know when to use photos from social media* [Homepage of The Guardian], [Online]. Available: www.journalism.co.uk/skills/how-to-know-when-to-use-photos-from-social-media/s7/a545631/ [2015, 2 November].

McLellan, T. 1999, "Fair game or fair go? Impact of news reporting on victims and survivors of traumatic events", *Asia Pacific Media Educator*, vol. 1, no. 7, pp. 53–73.

Meares, A. 2005, "How do you photograph a tragedy?", in *Shaking our foundations*, eds J. Park & J. Aagaard Jensen, International Federation of Journalists. Available: http://webarchive.unesco.org/20161111150756/http://portal.unesco.org/ci/en/files/18312/11092454711ifj_tsunamireport.pdf/ifj_tsunamireport.pdf [2018, 27 August].

Merrill, J. 2015, 26 August-last update, *Shoreham airshow: Medic at scene of crash says he saw shocked members of public who were filming on mobile phones* [Homepage of The Independent], [Online]. Available: www.independent.co.uk/news/uk/home-news/shoreham-airshow-medic-at-scene-of-crash-says-he-ran-past-laughing-and-gawping-members-of-public-who-10473379.html [2016, 7 January].

Moeller, S.D. 1999, *Compassion fatigue: How the media sell disease, famine, war and death*, Routledge, London.

Moore, S. 2011, 24 October-last update, *Yes he had it coming but did we have to gloat like barbarians* [Homepage of Mail Online], [Online]. Available: www.dailymail.co.uk/news/article-2052374/Gaddafi-dead-He-coming-did-really-need-gloat-like-barbarians.html [2011, 11 December].

Muller, D. 2013, "Black saturday bushfires and the question of consent", *Ethical Space*, vol. 10, no. 1, pp. 36–42.

Newton, J. & Duncan, S. 2012, "Journalists and the bereaved: Constructing a positive approach to the teaching of death reporting", *Journalism Education*, vol. 1, no. 2, pp. 59–67.

Nissenbaum, H. 2004, "Privacy as contextual integrity", *Washington Law Review*, vol. 79, no. 1, pp. 119–158.

Ofcom. 2005, 26 September-last update, *Ofcom broadcast bulletin 44* [Homepage of Ofcom], [Online]. Available: http://stakeholders.ofcom.org.uk/binaries/enforcement/broadcast-bulletins/pcb51/issue44.pdf [2015, 3 November].

Ofcom. 2017, April-last update, *Ofcom broadcasting code guidance* [Homepage of Ofcom], [Online]. Available: https://www.ofcom.org.uk/__data/assets/pdf_file/0005/100103/broadcast-code-april-2017.pdf [2018, 19 September].

Parkinson, H. 2015, 27 July-last update. *Twitter removes lifted jokes over copyright infringement claims* [Homepage of The Guardian], [Online]. Available: www.theguardian.com/technology/2015/jul/27/twitter-removes-lifted-jokes-copyright-infringment-claims [2018, 27 August].

PCC. 2008, 23 April-last update, *Kirkland v Wiltshire Gazette and Herald* [Homepage of Press Complaints Commission], [Online]. Available: www.pcc.org.uk/cases/adjudicated.html?article=NTA4Nw==&type= [2015, 15 October].

PCC. 2009, 29 September-last update, *Goble v people* [Homepage of Press Complaints Commission], [Online]. Available: www.pcc.org.uk/cases/adjudicated.html?article=NjA4MQ== [2015, 2 November].

PCC. 2010, 4 January-last update, *Deborah Rundle v Sunday Times* [Homepage of Press Complaints Commission], [Online]. Available: www.pcc.org.uk/cases/adjudicated.html?article=NjE1NQ== [2015, 15 October].

PCC. 2011, 8 February-last update, *Baskerville v Daily Mail* [Homepage of PCC], [Online]. Available: www.pcc.org.uk/cases/adjudicated.html?article=NjkzNA== [2015, 2 November].

PCC. 2012, 1 January-last update, *Editors code of practice* [Homepage of Press Complaints Commission], [Online]. Available: www.pcc.org.uk/assets/696/Code_of_Practice_2012_A4.pdf [2016, 15 April].

Peterson, C. 2010, *Losing face: An environmental analysis of privacy on Facebook*, MIT, Available at SSRN: http://ssrn.com/abstract=1550211 or http://dx.doi.org/10.2139/ssrn.1550211.

Phillips, W. 2011, "LOLing at tragedy: Facebook trolls, memorial pages and resistance to grief online", *First Monday*, vol. 16, no. 12. Available: http://journals.uic.edu/ojs/index.php/fm/article/view/3168 [2018, 2 July].

Phillips, W. 2013, "The house that fox built anonymous, spectacle, and cycles of amplification", *Television & New Media*, vol. 14, no. 6, pp. 494–509.

Robertson, G. & Nicol, A. 2007, *Media law*, 5th edn, Sweet & Maxwell, London.

Ronson, J. 2015, *So you've been publicly shamed*, Picador, Oxford..

Sargent, J. 2015, 12 January-last update, *Respecting the eyewitness: Learning lessons from the Paris shootings* [Homepage of First Draft], [Online]. Available: https://medium.com/1st-draft/respecting-the-eyewitness-learning-lessons-from-the-paris-shootings-a1391a7cbe23#.xllxvz5lc [2015, 2 November].

Satter, R. 2015, 11 January-last update, *Witness to Paris officer's death regrets video* [Homepage of Associated Press], [Online]. Available: http://bigstory.ap.org/article/5e1ee93021b941629186882f03f1bb79/ap-exclusive-witness-paris-officers-death-regrets-video [2015, 2 November].

SDNY 2013, *Agence France Presse v. Morel*, 934 F. Supp. 2d 547. Available: https://www.leagle.com/decision/infdco20130115975 [2018, 28 August].

Singer, S. & Turrell, E. 2006, "The 21st-century journalist", *Entertainment Law Review*, vol. 17, no. 2, p. 55.

Toor, A. 2016, 14 June-last update, *French terror suspect reportedly streamed attack on Facebook live* [Homepage of The Verge], [Online]. Available: www.theverge.com/2016/6/14/11930916/france-terrorist-larossi-abballa-facebook-live-video [2018, 27 April].

Townend, J. 2012, 9 October-last update, *A journalist calls: Reporting death in a post-Leveson climate* [Homepage of Dart Center for Journalism and Trauma], [Online]. Available: http://dartcenter.org/content/journalist-calls-reporting-death-in-post-leveson-climate#.Vct86Uv_Qds [2015, 12 August].

Townend, J. 2014, "Layers of consent", *Ethical Space*, vol. 11, no. 3, pp. 25–29.

Voorhoof, D. 2009, "European court of human rights case of TASZ v. Hungary", *IRIS*, no. 7, pp. 2–3.

Wahl-Jorgensen, K. & Pantti, M. 2013, "The ethics of global disaster reporting: Journalistic witnessing and the challenge to objectivity", in *Global media ethics: Problems and perspectives*, ed. S.J.A. Ward, Wiley-Blackwell, Chichester, pp. 191–213.

Wagner, K. 2018, 19 January-last update, *Justine Sacco, the PR exec who was fired from IAC for her tweets, has landed back at IAC's Match Group* [Recode.net home page], [Online]. Available: https://www.recode.net/2018/1/19/16911074/justine-sacco-iac-match-group-return-tweet [2018, 28 August].

Ward, S.J.A. 2005, "Philosophical foundations for global journalism ethics", *Journal of Mass Media Ethics*, vol. 20, no. 1, pp. 3–21.

Ward, S.J.A. 2008, "Global journalism ethics: Widening the conceptual base", *Global Media Journal*, vol. 1, pp. 137–149.

Watson, S.T. 2014, 20 November-last update, *Social media magnify power of breathtaking megastorm* [Homepage of Buffalo News], [Online]. Available: http://buffalonews.com/2014/11/20/social-media-magnify-power-of-breathtaking-megastorm/ [2018, 30 April].

Whittle, S. & Cooper, G. 2009, *Privacy, probity and public interest*, Reuters Institute for the Study of Journalism, Oxford.
Xalabarder, R. 2012, "Google news and copyright", in *Google and the law*, TMC Asser Press, The Netherlands, pp. 113–167.
Zelizer, B. 2010, *About to die*, Oxford University Press, Oxford.
Zimmer, M. 2010, " 'But the data is already public': On the ethics of research in Facebook", *Ethics and Information Technology*, vol. 12, no. 4, pp. 313–325.

9 Conclusion

As Emily Bell, director of the Tow Centre for Digital Journalism, describes it, the news ecosystem "has changed more dramatically in the past five years than perhaps at any time in the past five hundred" (Bell, 2016). A key part of this is the use of user-generated content—provided for free by untrained members of the public as opposed to paid journalists—which has already disrupted the journalistic field and economy. The consequences are clear in humanitarian reporting: the pressure to use more of such content and the increasing use of aid agency facilities by hard-up papers. But there were also disruptions from the view of struggles over cultural capital.

In the aftermath of the south Asian tsunami of 2004 and Hurricane Katrina in 2005, there were claims that the dynamics of humanitarian disaster reporting would be disrupted by the ability of citizens to share and publish their own accounts of events (Gillmor, 2005; Robinson and de Shano, 2011). The 2004 tsunami and Hurricane Katrina in 2005 both occurred in a pre-Twitter age when Facebook was still restricted to a small audience. Again however when social media came into play, it seemed that this was again handing more power to a public to share and publish content, and create the dominant images or text of an event while cutting out the need for this to be mediated via media organisations.

So it is easy to focus on the technological changes. Yet what is interesting is that, despite how easy it has become to publish and share content, much has stayed the same in covering humanitarian crises.

Of course, boundaries have blurred and (re)negotiation occurred. No serious journalist would operate today without a Twitter account; no aid agency would ignore Facebook. Any story about a dramatic, rapid-onset disaster would include some kind of user-generated content—whether video footage, livestreaming from an app like Instagram, Snapchat or Facebook Live, or even just a tweet/post to place oneself at the scene. The answer to the question of whether this technological revolution has led to a disruption in the journalistic field seems at first obvious.

Media organisations often use UGC because it allows the reporting (if not the journalist themselves) to take place at the centre of the drama; it gives "the whiff of authenticity" (Anderson, Coleman and Thumim, 2015:90).

As these researchers put it when reviewing studies into citizen journalism in local outlets in Philadelphia and Leeds: "The role of so-called citizen journalism in other words may be more to provide traditional outlets with authentic content than rewrite the rules of news production" (2015:90–91).

Yet, the technological ease of reporting faraway stories should not blind us to what remains the same. Humanitarian disasters may seem sudden and chaotic, but they have always been reported using preferred, ritualised strategies (Allan and Peters, 2015). These rituals may have been updated for a web 2.0 age—start a liveblog, search Twitter for first-hand experiences, scan Facebook for pictures and YouTube for video—but Bourdieu's concept of the 'rules of the game' are still understood by journalists today. The same is true of aid agencies who now operate 24-hour press offices, update their own Twitter feeds, adopt hashtags and aim to get pictures and words from the ground as soon as possible.

This book shows that there has been a control, co-option and cloning of user-generated content to ensure the significant players in the field retain their power. For, despite Robinson and de Shano's optimism that the "user is now the centre of the universe" (2011:977), the mainstream media still controls which stories we see and when.

The Source–Media Relationship

The relationship between journalists and aid agencies remains a powerful and symbiotic one. When journalists were questioned about disaster coverage they did not think UGC or UGC creators had replaced aid agencies as a major source, despite the amount of UGC they used. While editors saw UGC as vital in the immediate aftermath of disasters, before a journalist could get there, they still clung to the idea of the journalist as the primary witness. They also frequently used tweets from authority figures or other news organisations as a means of updating their liveblogs or websites. Reporters on the ground, meanwhile, often said that they had no time to incorporate user-generated content into their accounts because of technological difficulties, insufficient time to verify the content or doubts about its value.

While reluctance to use UGC was often framed as a respect for journalistic norms, several journalists admitted fearing for their own jobs or even the future of paid journalism if editors could get "perfectly good wobbly pictures" from a creator of UGC (Interviewee 36).

Journalists were often reluctant to admit the full extent of the support they received from aid agencies, compared with what the aid agencies themselves said they gave journalists. But they still described aid agencies both as important sources of attribution (Carlson, 2009) and, still, as gatekeepers into the world of disasters.

Despite all the new sources of information in the internet age, the combination of scarce resources and security worries meant that journalists were just as dependent, if not more so, on aid agencies as they had been before the worldwide web. The slashing of foreign budgets meant that for some cash-strapped media organisations, accepting trips or content was a cheap—or even the only affordable—way of covering some stories. The increased targeting of journalists by armed groups also meant that accepting aid agency support was seen as a sensible precaution.

The professionalisation of aid agency media teams meant that press officers were able to conform with what Waisbord (2011) calls a "journalistic logic". This, in turn, made it easier for some journalists to co-operate closely with aid agencies. Case studies were worked out in advance and press releases were written by former reporters, so the journalists had confidence that the aid agencies would supply what was needed for a news feature.

Even when the aid agency did not foot the whole bill for a trip, journalists often framed the stories they gathered in the way aid agencies wanted. One of the most interesting aspects of the aid agency/journalist relationship was the use of fixers (Murrell, 2010, 2013, 2014). Many aid agencies said that while they would not pay for a journalist's flights, they would set them up with translators and fixers. Often a press officer would act as a *de facto* fixer, negotiating with in-country staff for the journalist.

This is not to say that the journalist has no autonomy—as World Vision's disastrous trip to Sierra Leone with Jake Wallis Simons of the *Daily Telegraph* and actor and singer Elizabeth McGovern shows (Wallis Simons, 2013).[1] And it is not to say that aid agencies see themselves as powerful players in the field. Several agencies voiced the frustrations of trying to get media involved in the stories that they wanted to promote, or the experience of organising interviews, quotes or even trips that were then dropped at a moment's notice. While aid agencies could play a significant role, they felt the power dynamics still resided very much with the mainstream media.

Similarly, while aid agencies devoted a lot of time to thinking about user-generated content and knew they needed to be engaging in a social media world, their focus remained the mainstream media. As one senior humanitarian press officer pointed out, not only was the audience potentially much larger for a piece on the BBC's *Ten O'Clock News*, crucially the mainstream media gave the aid agency viewpoint 'validation' (Interviewee 13).

There were attempts to prioritise social, over mainstream, media—Oxfam GB's Twitter takeover and #hiddencrisis, for example. But these were individual projects rather than the rule.

Cloning, Co-Opting, Controlling Content

That UGC had had an influence on journalistic practice was beyond doubt. The increasing personalisation of news reporting was also undoubted, even if few journalists went as far as Anderson Cooper and Sanjay Gupta in what Chen sarcastically called the mutation from "CNN's Excellent Haiti Coverage to CNN's Excellent Haiti Adventure" (2010). It was no longer enough for reporters to have a presence on the printed page, website or evening news. They had to be permanently available to their audience on the medium that audience uses to communicate—social media.

This was often expressed by editors not in terms of enhancing the quality of the *journalism* but enhancing the *brand*—ensuring that traffic was driven to the news website or programmes. While editors realised this might be difficult for reporters in the extreme circumstances of a humanitarian disaster, journalists frequently felt under pressure to perform as a multimedia correspondent. The result, in the words of one foreign affairs editor at a commercial broadcaster, was that "there really isn't any downtime anymore" (Interviewee 40).

But by cloning the strengths of UGC, journalists could negate some of the challenges it brought to the field. The mainstream media has sought to capture and use user-generated content, taking ownership of it in a more organised way than simply mimicking typical UGC tropes. Just as the media sought to 'normalise' blogs (Singer, 2005), now media organisations have turned to popular forms of UGC, mainly pictures. But to maintain their authoritative position, they have used strategies to make sure such content is segregated or co-opted (Wahl-Jorgensen, 2015).

Thus, the BBC UGC Hub successfully promoted the idea of user-generated content as another source for the corporation's journalists, rather than a different (and potentially challenging) form of journalism altogether. The Hub's journalists had evolved to act as effective 'gatewatchers' (Bruns, 2008) and gatekeepers, both requesting content (via the 'Have Your Say' form) and choosing it (when there was breaking news). By setting up the Hub, the BBC was taking ownership of user-generated content, and the corporation's journalists mostly experienced UGC through the medium of the Hub.

The concern for the Hub's journalists was to verify such content to BBC standards—and to ensure they were ahead of their competitors. This meant sometimes putting a monetary value on some UGC—although it was frequently traditional journalistic skills—physically going out into the scene of the news story to find creators—that were needed to secure the content.

With the GuardianWitness app, journalists spoke of their dedication to giving opportunities to different voices and to telling different stories. Yet, the overall control of the stories and the pictures was something the staff found difficult to justify. There were also problems in reaching out to those at the centre of these stories. As a result, the content was more likely to be dominated by NGOs with their own particular agendas to explain than those who had experienced a crisis themselves; this was something that staff at GuardianWitness co-operated with and accepted in order to get the content they required.

NGOs had initially reacted with confusion and concern to the growth of user-generated content; many still had not worked out policies how to interact with it. The stated commitment of many aid agencies to ensure beneficiaries were heard and the degree to which UGC made this possible was something that many felt nervous about in reality; the focus was more often on satisfying their donors instead. There were attempts to facilitate content creation by beneficiaries, but these tended to be limited to individual projects (e.g. giving cameras to children at the Zaatari refugee camp), rather a regular commitment.

Aid agencies had been unsettled by the rapid introduction of UGC into the sphere of humanitarian reporting. They were entranced by the idea that sites like Twitter could be used for advocacy and fundraising, but frequently felt lost as to how they could best harness this power. They wanted to exploit it, but UGC as a free exchange of ideas was often not something aid agencies could deal with. They ended up trying to rigorously control or pay lip service to beneficiaries. There were notable exceptions—for example, the Oxfam GB Twitter Takeover of 2013 did seem a genuine attempt to allow beneficiaries to have a free voice, although it was still mediated by an Oxfam helper.

The use of the bloggers was a halfway house for aid agencies—to be engaged on social media, but in a controlled way. Bloggers' trips were planned as carefully as press trips and, for the aid agency, had the added benefit that the bloggers they chose were not experienced foreign reporters who might ask probing questions about their effectiveness. The bloggers also had audiences who were fertile ground for aid agency fundraisers.[2]

Some agencies thought the influence of bloggers was on the wane and were moving towards social media influencers on sites such as Instagram. There were mixed feelings here however. Some agencies felt that bloggers had more space and time to devote to important issues; others, while feeling that Instagram Stories was key were concerned that the commercialisation of the social media space was such that aid agencies could be reduced to "just another thing to flog", while the fact that more stories were emerging about influencers' social media postings from years earlier meant that unlike more media-literate journalists, associating with such influencers could lead to brand damage.

Blurred Boundaries—Working Between Media and Aid Agencies

As well as the boundary renegotiation over *what* constituted journalism, the question of *who* is a journalist went beyond the comparison between mainstream media and creators of UGC. Journalists spoke approvingly of press offices like Save the Children UK's, which had employed former journalists and which easily produced materials that conformed to, and suited, mainstream media norms.

Many aid agencies now employed significant numbers of former journalists in their offices. They were given job titles that mirrored those of journalists—such as 'head of World News Unit' or 'news editor'. And press offices' ability to provide content for journalists had undoubtedly improved with more accessible technology.

The best-funded agencies could now put their press officers on the ground and equip them with BGANs or Quicklinks so they could produce their own live broadcasts; the smaller ones were still relying on voice recorders and smartphones. This meant that there was an increasing divide between the larger, better-funded aid agencies and the smaller groups who found it difficult to make themselves heard. While some did punch above their weight—for example, Islamic Relief—those with smaller press offices found their voices squeezed out. A search of Lexis-Nexis of UK publications between the dates 1 January 2013 and 1 January 2018[3] found that, while Age International had ten stories where they were mentioned three times, in comparison Islamic Relief had 192, Christian Aid 2,013, and Save the Children and Oxfam both more than 3,000.

What also emerged was a blurring in a small but significant group of journalists, who moved between working for aid agencies and working for mainstream media organisations. That journalists change career to work for NGOs is not new; many of my interviewees had done just that. What was interesting was that many now moved back and forth between the two worlds.

This is not just a consequence of the social media age. Speaking at a debate organised by the Frontline Club as far back as 2008, the award-winning photographer Marcus Bleasdale said that 80–85% of his work was financed by humanitarian agencies (Frontline Club, 2008). That taking NGO-funded trips was now accepted practice was demonstrated by the ease with which journalists talked about it. For example, a Sunday broadsheet editor described how he commissioned a piece from a former journalist who was working for an aid agency, while a tabloid reporter said she had considered asking an NGO to finance her so she could cover the kind of stories she wanted without them becoming 'vanity' projects (Interviewee 20; Interviewee 34). Similarly, the aid agencies openly discussed employing freelancers—with some paying freelancers a day rate to cover a story or take photographs that they could then promote to a newspaper.

This raises the spectre of what Dan Gillmor has called 'almost-journalism', when advocates fund reporting, raising the question of what agenda is being followed. Gillmor has criticised NGOs and human rights organisations who have entered the media field: "They're falling short today in several areas, notably the one that comes hardest to advocates: fairness" (Gillmor, 2008).

Journalists, however, did not see any problem in 'swapping sides'; they identified themselves as journalists who just happened to work for a time with aid agencies. They felt they could demarcate the roles they played easily and inhabit the habitus required as and when necessary. In fact, they articulated more unease about being employed as journalists but having to 'embed' themselves with aid agencies (Cottle and Nolan, 2009) in order to fulfil a story for a media organisation.

The Virtual Doorstep

In the aftermath of the 7/7 bombings, the idea emerged of 'cameraphone ghouls' (Naughton, 2005), whose first reaction was to film the tunnels and the injured, after escaping serious injury themselves. A decade on, Tony Kemp, as mentioned earlier, the first medic at the scene of the aircrash in Shoreham, England, in August 2015, said those who did the same there were 'victims' too (Merrill, 2015).

A decade on however, capturing a video or photo, sending a tweet or posting on Facebook, has now become a reflex reaction to prove you were there—whether at the announcement of a new Pope,[4] a concert or the Nepal earthquake (Kanwal, 2015).

But this new ability of a citizen, to establish him or herself as a co-witness, is not the same as controlling how a story is framed.

While many of the UGC creators interviewed for this research did feel flattered and happy that their content had been used by a mainstream media organisation, many had no idea that a simple tweet or picture or video would be used in the way it was—and potentially open them up to criticism. Many were overwhelmed; they felt attacked and under pressure, and decided to go offline temporarily as a result.

The way that journalists approach creators of content has changed over the years from a 'smash and grab' approach (not asking permission) to a fairly cursory tweet asking if the content can be used. But, all too often, the focus is on obtaining copyright approval rather than engaging in a conversation with a possibly traumatised individual. Rarely, it seems, do journalists think about what it is like for an eyewitness to be repeatedly contacted on Twitter or Facebook after being caught up in a disaster.

Previous work on the practices of citizen journalists (in particular Reich, 2008; Robinson and De Shano, 2011) has tended to look at citizen journalists who commit to a particular event or website for some time. But we also need to focus on the 'accidental journalists', sourcing and how creators of UGC perceive themselves in relation to journalists, examining qualitatively how they feel about the use of their work and their interactions with journalists. Most of those who circulated pictures or wrote tweets that were picked up by the mainstream media did so only in the immediate aftermath of a disaster; they did not go on to question the mainstream media's version of events. Unlike in the aftermath of 2005's Hurricane Katrina (Robinson, 2009), there was no attempt by most of the creators to (re)vision the coverage. While many of those asked expressed normative values—that they felt a duty to report, and that they would do the same thing again if caught up in another disaster, it was clear that many had felt hurt and worried about the use of their content.

Distant Suffering

> Of all the things I regret more than anything is that it never crossed my mind to interview anyone who didn't speak English because I knew they wouldn't use it . . . and so consequently all the victims, all the people who the story was about were voiceless.
>
> (Michael Buerk, reflecting on Ethiopia, 1984, Al Jazeera English, 2009)[5]

When Michael Buerk revealed to the world the extent of the famine in Korem, Ethiopia, in 1984, his report became the biggest story that the BBC covered in the 1980s until the fall of the Berlin Wall (Simpson, 1998). Yet his seven-minute package that led the *Six O'Clock News* contained only his own voice and that of a white doctor (Cooper, 2007).

Even Buerk says he would not cover the story that way now; the Ethiopian famine proved a turning point, not just in the scale of disaster coverage, but in how both media and aid agencies should move away from the Western-centric approach that allowed survivors of a crisis to be viewed but not to speak (Coulter, 1989; Coulter and Cooper; 2015).

What Sambrook calls the "narrative of conscience" (2010:6), or Bell's "journalism of attachment" (1997), was very much mediated witnessing, a central part of journalism's claim to authority (Zelizer, 2007). But, as Chouliaraki points out, the intervention of "ordinary witnessing" gives a different "epistemology of authenticity that relativises the empiricism of facts in television news by placing it side by side with the empiricism of emotion" (2010:308).

What has changed? It would be wrong to say that there was no difference; the journalists certainly felt that UGC was now a key part of their coverage, and that getting the key eyewitness picture, tweet or video that had been shared by someone caught up in a disaster was one of the most important things to do.

But most of the time the content created by ordinary people was seen very much as just another source rather than a new way of reporting. To return to Carlson's (2009) imagery of a journalist's relationship with their source as duelling, dancing or dominating, this was domination. The way that journalists maintained their status in the journalistic field was to use UGC as a technologically sophisticated version of the case study. Even when this challenged reporting, it was seen as a temporary fix until the journalist could get there.

Yet did these first-person, subjective accounts help to narrow the distance between viewer and survivor? Up to a point, yes. The audiences themselves responded positively to such material (Wardle and Williams, 2008; Wahl-Jorgensen, Williams and Wardle, 2010), while the journalists often felt it provided authenticity and diversity. Interviewee 38, an international editor for a commercial broadcaster, added that being on Twitter meant he had been pointed to UGC on Twitter by people hoping to alert him to problems they were facing during a humanitarian crisis.

But because this material was mediated using the traditional journalistic lens, distance was not erased. This simulated closeness frequently descended into the banal (a tweet or an Instagram snap), leaning towards a "commodified sentimentalism that reduces witnessing to voyeurism" (Chouliaraki, 2010:306).

Aid agencies continued to mediate the voices of survivors and beneficiaries, despite their mandate to make those voices heard. Most, for reasons they articulated as practical, wanted to continue to get beneficiary voices to a wider public in the same way as before—by interviewing them themselves and passing on quotes/pieces to journalists. While most of them paid lip service to the idea, many questioned how giving beneficiaries access to Twitter or cameras would actually be beneficial or overcome distance in the long run.

When aid agencies did attempt a different way, they chose to communicate their message through bloggers and then social media influencers rather than beneficiaries. This meant that the emotional focus became the bloggers' shock, rather than the beneficiaries themselves, "appropriating the sufferer as someone who shares our humanity" (Chouliaraki, 2010:306). So while there have been pockets of progress, it is impossible to say that there has been a major shift in distance.

Tavernor's work on "transient compassion" (2017) is perhaps most apt here. She disagrees with Moeller's reading of compassion fatigue and argues that shocking images do not result in a lack of public response. Instead, the problem is that, while the public does respond, the response is fleeting and insubstantial; it may be nothing more significant than liking or sharing a Facebook page. This is something that aid agencies themselves are grappling to overcome, so that communicating suffering through UGC and social media does not descend into nothing more than a "guilt-assuaging pastime" (Miller, 2017:263).[6]

The Future

This particular study was carried out at a time of rapid change in news technologies; as such it has to represent a snapshot in time but provides a valuable point to consider these unresolved issues.

While user-generated content is now an established part of the coverage of a crisis, then it was still filtered through the mainstream media. This, of course, is changing, with many people turning to Facebook, Google and Twitter to get their news, and the introduction of Instant Articles by Facebook, Snapchat's Discover channels, Twitter's Moments, Apple News and Google's Accelerated Mobile Pages.[7] A newspaper's pages or a broadcaster's programme schedule has been replaced by a website's algorithm, which may be even less transparent.

Sites like BuzzFeed and VICE News have diversified into news as well,[8] and their presentation style is influential. It would be interesting to examine whether these webnative sites use UGC in the same way as mainstream media, and whether their relationship with UGC creators is similar.

Yet BuzzFeed's decision to employ journalists from legacy media such as the *Guardian* suggests that, while sites like this claim to challenge newsroom norms, they too may be starting to adopt mainstream media practices to ensure credibility.

Media organisations like the BBC do now have guidelines for dealing with creators of UGC—for example, trying to avoid multiple approaches (BBC, 2015)—while organisations like EyeWitness Media Hub have attempted to provoke ethical debates on the subject (Wardle, 2015). Yet it still remains a pertinent issue—in the aftermath of the March 2016 Brussels terror attacks, people who found themselves caught up in it were pursued by journalists (Belam, 2016). One, David Crunelle, even wrote his own blogpost describing how he was hounded by media organisations from around the world, even after he had licensed CNN to use his Vine video as an exclusive (Crunelle, 2016, translated by Dubberley).

Second, the blurring of boundaries—examining those journalists who fund their career as reporters by accepting short-term contracts with aid agencies or by working as consultants for NGOs while still freelancing—is also an area which would warrant more investigation, a subset of journalists who manage to inhabit two different fields, and absorb and maintain two different forms of habitus.

That UGC has had an effect is undoubted, both in the switch from product to 'process' journalism (Jarvis, 2009) and in the aesthetics that are now frequently employed. The BBC's use of the Periscope app after the Nepal earthquake to record devastation in a Sindhupalchok village owed a huge aesthetic debt to UGC (Cooper, 2015). But the interviews also reveal how journalists have attempted to continue their primacy in arenas where contestation occurs by cloning characteristics of UGC (the constant updates via Twitter, the personal subjective approach) in an attempt to adopt the strengths that UGC has brought to communication of humanitarian disasters and maintain their own supremacy.

Any analysis of humanitarian reporting must include the role that the aid agencies play—a gatekeeping role that has been challenged by the advent of UGC. This book reveals that despite the claims for UGC, the source–media relationship between aid agencies and journalists on the ground remains a pervasive one. By questioning journalists from the major national print and broadcast organisations and representatives of the Disasters Emergency Committee, it provides an opportunity to compare and contrast how both sides articulate norms and practices during crisis situations. It shows up the divergences for example in what journalists say they will accept from aid agencies on the ground, and aid agencies' accounts.

As aid agencies have tried to adapt to a new media world, this has not necessarily enabled their beneficiaries to have more of a voice. Agencies have tried to best utilise social media and user-generated content but for the smaller agencies, this has not provided a more level playing field when it comes to media attention and fundraising opportunities.

Most research into the use of UGC has traditionally focused on the effects on journalists or the audience exposed to such content; this however attempts to triangulate considerations of how the dynamics of humanitarian disasters are being shifted by addressing the creators as well. It highlights how issues around privacy and permissions have still not been resolved.

While many UGC creators were happy that their content had been taken without permission, for others it was a distressing and traumatic experience to find that their content had spread far further than they had ever intended it to go. It raises ethical questions about how such creators should be treated, and whether journalistic codes should be updated to reflect the pursuit of individuals in cyberspace as well as in real life.

It reveals also the fact that a small minority are attempting to reframe the narrative by using their content not just to bear witness but to participate in what have been called 'acts of journalism' (Myers, 2014) and to create a historical record that challenges the journalistic narrative particularly in 'Quakebook' which was codified as a deliberative and normative attempt to resolve moral issues around responses to the earthquake.

Of particular interest are the ethical and legal dilemmas that this has thrown up concerning permissions, privacy and payment. The basics of consent—even with no monetary recompense—still appears to be a fraught area and, while privacy concerns no longer mean that putting anything online is apparently fair game, much still rests on an individual's or an organisation's normative and deontological approaches. With UGC now an established part of any disaster coverage, such debates need to be resolved.

Notes

1. Wallis Simons' (2013) account of actor Elizabeth McGovern's trip to Sierra Leone with World Vision included the revelations that McGovern mixed up Darfur and Dakar, was unaware that World Vision was a faith-based organisation, that the charity had paid her band £28,000 to be the charity's ambassadors and that she thought African people had sex more freely than in the West.
2. Liz Scarff who created the #blogladesh trip for Save the Children UK referred to it as targeting a niche community, adding: "[Social media] operates in a sense like local and regional newspapers used to. . . . Of course you can't replicate exactly what we did—because it was a novelty—you can't do mummy bloggers again but you could do food bloggers or political bloggers" (Cooper, 2011:34).
3. This was before media coverage of the humanitarian scandals which hit Oxfam and Save the Children in 2018.
4. In 2013, NBC's *Today* show posted on Instagram two pictures comparing the announcement of Pope Benedict in 2005 and the announcement of Pope Francis in 2013. The second picture shows the surge in cameraphones in less than a decade www.instagram.com/p/W2BuMLQLRB/.

5. In 2009, Buerk took part in Al Jazeera English's *The Listening Post* programme, looking back on 25 years since his reports on the Ethiopian famine. His remarks can be found at 6'27 onwards www.youtube.com/watch?v=fkxGFC1zcxQ.
6. Two campaigns encapsulate this. The 2013 'Liking Isn't Helping' for Crisis Relief Singapore followed an earlier campaign by Unicef Sweden entitled 'Likes Don't Save Lives' www.youtube.com/watch?v=2_M0SDk3ZaM.
7. See Bell (2016) and Hazard Owen (2016) for a brief description of these.
8. Buzzfeed notably collaborated with BBC News over two major investigations relating to the charity Kids Company and match-fixing in tennis (Burne, James and Reynolds, 2016) and has hired journalists from the *Financial Times,* the *Guardian* and *Los Angeles Times*. VICE News, launched in March 2014, won two Peabody Awards in 2015 for its films inside so-called Islamic State and a Chicago school for at-risk youth (Mojica, 2015), and was nominated for four Emmys the same year. Snapchat launched Snapchat Discover in January 2015, which has daily feeds from media companies such as the *Wall Street Journal*, CNN and the *Daily Mail* (Flynn, 2016).

References

Allan, S. & Peters, C. 2015, "Visual truths of citizen reportage: Four research problematics", *Information, Communication & Society*, vol. 18, no. 11, pp. 1348–1361.

Anderson, C.W., Coleman, S. & Thumim, N. 2015, "How news travels: A comparative study of local media ecosystems in Leeds (UK) and Philadelphia (US)", in *Local journalism: The decline of newspapers and the rise of digital media.*, ed. R.K. Nielsen, I B Tauris, London, pp. 73–93.

BBC. 2015, 26 March-last update, *Social media guidance for staff* [Homepage of BBC], [Online]. Available: http://news.bbc.co.uk/1/shared/bsp/hi/pdfs/26_03_15_bbc_news_group_social_media_guidance.pdf [2016, 1 April].

Belam, M. 2016, 24 March-last update, *Using Twitter and Facebook images of tragedies raises ethical dilemmas* [Homepage of The Guardian], [Online]. Available: www.theguardian.com/media/2016/mar/24/twitter-facebook-images-ethical-dilemmas-mobile-social-media [2016, 1 April].

Bell, E. 2016, 7 March-last update, *Facebook is eating the world* [Homepage of Columbia Journalism Review], [Online]. Available: www.cjr.org/analysis/facebook_and_media.php [2016, 27 March].

Bell, M. 1997, "TV news: how far should we go?" *British Journalism Review*, vol. 8, no. 1, pp. 7–16.

Bruns, A. 2008, *3.1. The active audience: Transforming journalism from gatekeeping to gatewatching*. Available: http://snurb.info/files/The%20Active%20Audience.pdf [2018, 2 July].

Burne James, S. & Reynolds, A. 2016, 19 January-last update, *Curious bedfellows? BBC-BuzzFeed news partnership remains casual despite major scoops* [Homepage of PR Week], [Online]. Available: www.prweek.com/article/1379834/curious-bedfellows-bbc-buzzfeed-news-partnership-remains-casual-despite-major-scoops [2016, 29 March].

Carlson, M. 2009, "Dueling, dancing, or dominating? Journalists and their sources", *Sociology Compass*, vol. 3, no. 4, pp. 526–542.

Chen, A. 2010, 18 January-last update, *Anderson Cooper saves boy as Haiti coverage reaches strange apotheosis'* [Homepage of Gawker.com], [Online]. Available: http://gawker.com/5451459/anderson-cooper-saves-boy-as-cnns-haiti-coverage-reaches-strange-apotheosis [2012, 19 May].

Chouliaraki, L. 2010, "Ordinary witnessing in post-television news: Towards a new moral imagination", *Self-Mediation: Citizenship and New Media, Special Issue Critical Discourse Studies*, vol. 7, no. 3, pp. 305–319.

Cooper, G. 2007, *Anyone here survived a wave, speak English and got a mobile? Aid agencies, the media and reporting disasters since the tsunami*, Nuffield College, Oxford.

Cooper, G. 2011, *From their own correspondent? New media and the changes in disaster coverage: Lessons to be learned*, Reuters Institute for the Study of Journalism, Oxford.

Cooper, G. 2015, 11 May-last update, *Nepal: Why are some disasters bigger than others?* [Homepage of BBC College of Journalism], [Online]. Available: www.bbc.co.uk/blogs/collegeofjournalism/entries/5e790c00-7ea9-4ee6-948d-3fee77cb26ce [2018, 28 April].

Cottle, S. & Nolan, D. 2009, 16 November-last update, *How the media's codes and rules influence the way NGOs work* [Homepage of Nieman Journalism Lab], [Online]. Available: www.niemanlab.org/2009/11/simon-cottle-and-david-nolan-how-the-medias-codes-and-rules-influence-the-ways-ngos-work/ [2015, 18 August].

Coulter, P. 1989, April-last update, *Pretty as a picture* [Homepage of New Internationalist], [Online]. Available: http://newint.org/features/1989/04/05/pretty/ [2015, 25 January].

Coulter, P. & Cooper, G. 2015, "NGOs, media and public understanding: 25 years on, an interview with Paddy Coulter, former head of media at Oxfam", in *Humanitarianism communications and change*, eds. S. Cottle & G. Cooper, Peter Lang, New York, pp. 79–91.

Crunelle, D. 2016, 30 March-last update, *The art of being in the wrong place at the right time: Behind the scenes of social media newsgathering* [Homepage of Eyewitness Media Hub], [Online]. Available: https://medium.com/@emhub/the-art-of-being-in-the-wrong-place-at-the-right-time-behind-the-scenes-of-social-media-3ee558630e93#.1x9uk8pfk [2016, 1 April].

Flynn, K. 2016, 27 January-last update, *Snapchat discover one year later: How 23 media companies are building stories for Evan Spiegel* [Homepage of IB Times], [Online]. Available: www.ibtimes.com/snapchat-discover-one-year-later-how-23-media-companies-are-building-stories-evan-2281851 [2016, 29 March].

Frontline Club. 2008, 28 February-last update, *The news carers: Are aid groups doing too much real newsgathering?* [Homepage of Frontline Club], [Online]. Available: www.frontlineclub.com/the_news_carers_are_aid_groups_doing_too_much_real_newsgathering_-_new_york_-_fully_booked/ [2016, 26 March].

Gillmor, D. 2005, 6 January-last update, *Tsunami and citizen journalism's first draft* [Homepage of Dan Gillmor], [Online]. Available: http://dangillmor.typepad.com/dan_gillmor_on_grassroots/2005/01/tsunami_and_cit.html [2012, 16 May].

Gillmor, D. 2008, 23 July-last update, *Helping the almost journalists do journalism* [Homepage of Center for Citizen Media Blog], [Online]. Available: http://citmedia.org/blog/2008/07/23/helping-the-almost-journalists-do-journalism/ [2016, 26 March].

Hazard Owen, L. 2016, 17 February-last update, *Facebook instant articles will open up to all publishers in April* [Homepage of Nieman Lab], [Online]. Available: www.niemanlab.org/2016/02/any-publisher-can-start-using-facebook-instant-articles-on-april-12/ [2016, 27 March].

Jarvis, J. 2009, 7 June-last update, *Product vs process journalism* [Homepage of Buzzmachine.com], [Online]. Available: http://buzzmachine.com/2009/06/07/processjournalism/ [2018, 9 May].

Kanwal, R. 2015, 2 May-last update, *Disaster selfies, treasure scavengers and saviours from the sky: Stories of heroism and tragedy in the aftermath of Nepal's killer earthquake* [Homepage of Mail Online], [Online]. Available: www.dailymail.co.uk/indiahome/indianews/article-3064702/Disaster-selfies-tourist sturned-reliefworkers-saviours-sky-Stories-heroism-tragedy-aftermath-Nepal-s-killerearthquake.html [2018, 11 June].

Merrill, J. 2015, 26 August-last update, *Shoreham airshow: Medic at scene of crash says he saw shocked members of public who were filming on mobile phones* [Homepage of The Independent], [Online]. Available: www.independent.co.uk/news/uk/home-news/shoreham-airshow-medic-at-scene-of-crash-says-he-ran-past-laughing-and-gawping-members-of-public-who-10473379.html [2016, 7 January].

Miller, V. 2017, "Phatic culture and the status quo reconsidering the purpose of social media activism", *Convergence: The International Journal of Research into New Media Technologies*, vol. 23, no. 3, pp. 251–269.

Mojica, J. 2015, 20 April-last update, *VICE News wins two Peabody awards* [Homepage of VICE.com], [Online]. Available: https://news.vice.com/article/vice-news-wins-two-peabody-awards [2016, 29 March].

Murrell, C. 2010, "Baghdad bureaux: An exploration of the interconnected world of fixers and correspondents at the BBC and CNN", *Media, War & Conflict*, vol. 3, no. 2, pp. 125–137.

Murrell, C. 2013, "International fixers: Cultural interpreters or 'People Like Us'?", *Ethical Space*, vol. 10, no. 2/3, pp. 72–79.

Murrell, C. 2014, *Foreign correspondents and international newsgathering: The role of fixers*, Routledge, New York.

Myers, S. 2014, 24 November-last update, *Why the man who tweeted the Osama bin Laden raid is a citizen journalist* [Homepage of Poynter], [Online]. Available: www.poynter.org/news/mediawire/131135/why-the-man-who-tweeted-bin-laden-raid-is-a-citizen-journalist/ [2015, 12 August].

Naughton, J. 2005, 17 July-last update, *Why I have serious doubts about the 'citizen reporters'* [Homepage of The Observer], [Online]. Available: www.theguardian.com/technology/2005/jul/17/comment.mobilephones [2016, 26 March].

Reich, Z. 2008, "How citizens create news stories: The 'news access' problem reversed", *Journalism Studies*, vol. 9, no. 5, pp. 739–758.

Robinson, S. 2009, " 'If you had been with us': Mainstream press and citizen journalists jockey for authority over the collective memory of Hurricane Katrina", *New Media and Society*, vol. Aug, no. 11, pp. 795–814.

Robinson, S. & De Shano, C. 2011, " 'Anyone can know': Citizen journalism and the interpretive community of the mainstream press", *Journalism: Theory, Practice, Criticism*, vol. 12, no. 8, pp. 963–982.

Sambrook, R. & University of Oxford. 2010, *Are foreign correspondents redundant? The changing face of international news*, Reuters Institute for the Study of Journalism, Oxford.

Simpson, J. 1998, *Strange places, questionable people*, Palgrave Macmillan, London.

Singer, J.B. 2005, "The political j-blogger 'normalizing'a new media form to fit old norms and practices", *Journalism*, vol. 6, no. 2, pp. 173–198.

Tavernor, R.M. 2017, *Communicating solidarity: The cultural politics and practices of humanitarian NGO campaigns*, Doctoral Thesis (PhD) edn, University of Sussex.

Wahl-Jorgensen, K. 2015, "Resisting epistemologies of user-generated content? Cooptation, segregation and the boundaries of journalism", in *Boundaries of journalism: Professionalism, practices and participation*, eds. M. Carlson & S.C. Lewis, Routledge, London and New York.

Wahl-Jorgensen, K., Williams, A. & Wardle, C. 2010, "Audience views on user-generated content: Exploring the value of news from the bottom up", *Northern Lights: Film & Media Studies Yearbook*, vol. 8, no. 1, pp. 177–194.

Waisbord, S. 2011, "Can NGOs change the news?" *International Journal of Communication*, vol. 5, pp. 142–165.

Wallis Simons, J. 2013, 22 December-last update, *Downton Abbey: Elizabeth McGovern's African adventure* [Homepage of Telegraph Media Group], [Online]. Available: www.telegraph.co.uk/culture/tvandradio/downton-abbey/10526655/Downton-Abbey-Elizabeth-McGoverns-African-adventure.html [2014, 19 February].

Wardle, C. 2015, 16 November-last update, *Ethical considerations for eyewitness media from EMHub's Claire Wardle* [Homepage of First Draft News], [Online]. Available: http://firstdraftnews.com/resource/ethical-considerations-for-eyewitness-media-from-emhubs-claire-wardle/ [2016, 1 April].

Wardle, C. & Williams, A. 2008, "ugc@BBC:Understanding its impact on contributors, non-contributors and BBC news", *BBC*. Available: www.bbc.co.uk/blogs/knowledgeexchange/cardiffone.pdf [2018, 28 August].

Zelizer, B. 2007, "On 'having been there': 'Eyewitnessing' as a journalistic key word", *Critical Studies in Media Communication*, vol. 24, no. 5, pp. 408–428.

Appendix
Interviewees

1. Aid Agencies

Job title	*Location*	*Date*	*Interviewee No*
Head of Media Relations	London	2 April 2013	1
Head of News	London	2 April 2013	2
Communications Manager	London	4 December 2013	3
Digital Manager	London	4 December 2013	4
Senior Press Officer	London	23 May 2013	5
Head of World News Unit	London	19 February 2013	6
World News Officer	London	19 February 2013	7
Head of Media	London	12 February 2013	8
Acting Head of Media	London	28 February 2013	9
Media and Advocacy Manager	London	16 April 2013	10
Media Manager	London	16 April 2013	11
Digital Manager	London	16 April 2013	12
Senior Humanitarian Press Officer	Oxford	26 February 2013	13
News Editor	London	28 February 2013	14
Acting Head of News	London	11 July 2013	15
Digital and Social Media Manager	London	11 July 2013	16
Head of Media and Corporate Communications	London	20 February 2013	17
Media Relations Manager	London	23 April 2013	18
Head of Communications	Via phone	7 May 2015	100
Director of News	Via phone	23 Sept 2013	101
Social Media Manager	Via phone	8 May 2018	102

2. Journalists

Print Journalists

Organisation	*Job Title*	*Location*	*Date*	*Interviewee Number*
Daily broadsheet	Foreign Editor	London	18 April 2013	19
Sunday broadsheet	Foreign Editor	London	19 February 2013	20
Daily broadsheet	Foreign Editor	London	17 April 2013	21
Sunday broadsheet	Deputy Foreign Editor/Foreign Correspondent	Via phone	3 April 2013	22
Sunday broadsheet	Former Foreign editor	Via phone	1 July 2015	23
Daily broadsheet	Foreign Editor	London	31 October 2013	24
Daily broadsheet	Foreign Correspondent	Via Skype	16 January 2014	25
Daily broadsheet	Foreign Correspondent	Via phone	2 April 2013	26
Sunday broadsheet	Foreign Editor	London	27 March 2013	27
Sunday broadsheet	Reporter	London	9 April 2013	28
Daily broadsheet	Foreign News Editor	London	2 May 2013	29
Daily and Sunday broadsheet	Foreign Correspondent for sister papers	Via phone	23 May 2013	30
Daily midmarket paper	Foreign Correspondent	London	10 April 2013	31
Daily midmarket paper	Former Foreign Editor	Via email	19 May 2015	32
Daily tabloid	Foreign Correspondent	London	7 April 2013	33
Daily Tabloid	Foreign Correspondent	Via email	18 May 2015	34

Broadcast/Online Journalists

Organisation	*Job Title*	*Location*	*Date of Interview*	*Interviewee Number*
Publicly owned, commercially funded	Editor	London	22 May 2013	35
Publicly owned, commercially funded	International Editor	Via phone	13 February 2013	36

Organisation	Job Title	Location	Date of Interview	Interviewee Number
Commercial broadcaster (terrestrial)	Assistant Editor (newsgathering)	London	19 March 2013	37
Commercial broadcaster (terrestrial)	International Editor	Via phone	14 February 2013	38
Commercial broadcaster (cable)	Head of International News	London	10 April 2013	39
Commercial broadcaster (cable)	Foreign Affairs Editor	London	13 February 2013	40
State funded broadcaster (cable)	Senior Officer Social Media	Via Skype	16 January 2014	41
State funded broadcaster (cable)	Project Manager Social Media	Via Skype	16 January 2014	42

3. UGC Creators

Job title	UGC created	Where used	Date of interview	Interviewee Number
IT manager	Tweet	BBC livepage	12 August 2015	43
English teacher	Audioboo	Guardian liveblog	3 & 4 June 2013; 21 May 2014	44
IT software designer	'Below the line'	Guardian liveblog	28 May 2013	45
Digital producer	Tweets	Guardian liveblog	8 May 2013	46
English teacher	'Below the line'	Guardian liveblog	8, 9, 10 May 2013	47
Music writer	'Below the line'	Guardian liveblog	8 May 2013	48
English teacher	'Below the line'	Guardian liveblog	27 May 2013	49
Engineer and writer	Tweet	Guardian liveblog	13 May 2013	50
Teacher	Have Your Say	BBC livepage	3 May 2013	51
Policy specialist	Have Your Say	BBC Livepage	7 November 2014	52
Welsh radio station worker	Tweet	BBC Livepage	31 July 2014	53
Student	Tweet	BBC Livepage	17 June 2014	54

Job title	*UGC created*	*Where used*	*Date of interview*	*Interviewee Number*
Small business owner	Tweet	BBC Livepage	19 June 2014	55
Technology company	Tweet	BBC Livepage	1 July 2014	56
American journalist	Tweet	BBC Livepage	10 June 2014	57
Tourist industry	Tweet	BBC Livepage	1 August 2014	58
Professor	Tweet	BBC Livepage	7 November 2014	59
Civil servant	Tweet	BBC Livepage	9 November 2014	60
Teacher	Have Your Say	BBC Live page	11 December 2014	61
Actor	Have Your Say and also picture via Twitter	BBC Livepage	8 December 2014	62
Journalist	Twitter	BBC Livepage	11 Nov 2014	63
Trader	Twitter	BBC Livepage	16 April 2015	64
Security Guard	Twitter (Vine Videos)	Multiple outlets	26 & 27 May 2013	65
Financial controller	Twitter (pictures)	Multiple outlets	30 January 2014	66
Video games designer	Twitter & YouTube (picture and video)	Multiple outlets	16 & 18 August 2015	67
Web designer and marketer	Blog	n/a	3 December 2015 & 14 January 2016	68
Head of marketing for publishing house	Have Your Say and YouTube video	Used on news channels	14 & 18 June 2013	69

4. Microethnography at GuardianWitness, the *Guardian*, London April/May 2013

Job Title	*Section*	*Date of Interview*	*Interviewee Number*
Editor	Liveblogs	29 April 2013	70
Community Manager	Guardian.co.uk	3 May 2013	71
Social and Communities Editor	Guardian Online	7 May 2013	72

Job Title	*Section*	*Date of Interview*	*Interviewee Number*
Reporter	Liveblogs	22 May 2013	73
Data Journalism	Guardian Online	7 May 2013	74
Community Co-ordinator	GuardianWitness	1 May 2013	75
Deputy Editor	Global Development	2 May 2013	76
Community Co-ordinator	GuardianWitness	1 May 2013	77
Community Co-ordinator	Global Development	30 April 2013	78
Network Production Editor	Guardian Online	1 May 2013	79
Night Editor	Guardian Online	29 April 2013	80
Multimedia Production Editor	Guardian Online	30 April 2013	81

5. Microethnography at UGC Hub, BBC, London June 2013

Job Title	*Date of Interview*	*Date of Interview*
Assistant Editor	11 June 2013	82
UGC Producer	10 June 2013	83
Senior Broadcast Journalist	12 June 2013	84
Assistant Editor, Social News	14 June 2013	85
UGC producer	10 June 2013	86
Journalist	13 June 2013	87
Senior Broadcast Journalist	11 June 2013	88
World Affairs Producer	1 September 2015 (via phone)	98
Presenter	8 September 2015 (via phone)	99

6. Microethnography at Save the Children UK, London

Job Title	*Date of Interview*	*Date of Interview*
Senior Campaigns Advisor	10 July 2013	89
Head of Government Relations	11 July 2013	90
Global Media Manager	11 July 2013	91
Senior Humanitarian Affairs Adviser	9 July 2013	92

Job Title	*Date of Interview*	*Date of Interview*
Senior Conflict & Humanitarian Advocacy Adviser	10 July 2013	93
Head of Global Media Unit	11 July 2013	94
Head of Digital	20 August 2013	95
Humanitarian Information and Communications Project Manager	9 July 2013	96
Digital Fundraising Manager	20 August 2013	97

Index

For Product Safety Concerns and Information please contact our EU representative GPSR@taylorandfrancis.com
Taylor & Francis Verlag GmbH, Kaufingerstraße 24, 80331 München, Germany

www.ingramcontent.com/pod-product-compliance
Ingram Content Group UK Ltd.
Pitfield, Milton Keynes, MK11 3LW, UK
UKHW022001180926
475701UK00003B/135
* 9 7 8 0 3 6 7 5 8 4 1 9 1 *